HADRON INTERACTIONS

The diagram on the cover shows a rare type of event observed by the UA1 detector at the proton–antiproton Collider at CERN (courtesy of the UA1 collaboration). The proton and antiproton collide head-on with a total energy of 540 GeV. Only particle tracks in the UA1 central detector with momentum components perpendicular to the beams of greater than 1 GeV/c are shown. The event shows the production of two hadronic jets, together with the rare occurrence of an 'isolated' muon. Accumulation of events of this type may yield evidence for the top quark (see chapter 13).

GRADUATE STUDENT SERIES IN PHYSICS

Series Editor: Professor Douglas F Brewer, M.A., D.Phil.
Professor of Experimental Physics, University of Sussex

HADRON
INTERACTIONS

P D B COLLINS

Reader in Theoretical Physics
University of Durham

A D MARTIN

Professor of Theoretical Physics
University of Durham

with a preface by

R H DALITZ

Royal Society Research Professor in Theoretical Physics
University of Oxford

ADAM HILGER LTD, BRISTOL
Published in association with the University of Sussex Press

British Library Cataloguing in Publication Data

Collins, P.D.B.
  Hadron interactions.—(Graduate student
  series in physics, ISSN 0261-7242)
  1. Hadrons
  I. Title   II. Martin, A.D.   III. Series
  539.7'216   QC793.5.H32

  ISBN 0-85274-768-3

Consultant Editor: **Professor R H Dalitz,** Department of Theoretical Physics, University of Oxford

Published by Adam Hilger Ltd, Techno House, Redcliffe Way, Bristol BS1 6NX, in association with the University of Sussex Press

Printed in Great Britain by J W Arrowsmith Ltd, Bristol

PREFACE

Great progress has been made in elementary particle physics over the past decade, both experimentally and theoretically. Following the experiments on the ψ/J particle at Brookhaven, New York (BNL) and on the ψ/J and ψ' particles with the electron–positron storage rings, SPEAR at Stanford, California (SLAC) and DORIS at Hamburg (DESY Laboratory), the quark model for mesons and baryons has become generally accepted. These experiments led us to recognise the existence of a fourth quark, the charm quark c, in addition to the quarks u, d and s, already indicated by the earlier hadron spectroscopy. More recent experiments, at FNAL near Chicago and at the electron–positron storage rings CESR at Cornell University, PEP at SLAC and PETRA at DESY, have established the existence of a heavier quark, the bottom quark b with mass about 4.5 GeV. There is a general expectation that at least one further quark, already named the top quark, t, will yet be found, and still heavier quarks might well exist, of course.

On the theoretical side, we now recognise a new hadronic degree of freedom termed 'colour'. This gives rise to an $SU(3)_C$ gauge symmetry for the interactions between quarks, the gauge field quanta being termed 'gluons'. The gluons are vector particles just like the photon, the quantum of the electromagnetic gauge field. The theory is known as quantum chromodynamics (QCD) and is believed to imply confinement for a quark or a gluon, or any hadronic system which is not a colour-singlet, although this has not been demonstrated rigorously. Further, QCD has the desirable property of being renormalisable. Thus, for the first time in history, theoreticians have a field theory of hadronic interactions which is finite, calculable in principle, and in qualitative concord with all of the experimental indications.

Also, a unification of weak and electromagnetic interactions has been achieved through an electroweak gauge theory, the 'standard' $SU(2) \times U(1)$ model, whose gauge bosons are the photon and the three weak bosons, $W^\pm$ and Z^0. This is also a renormalisable theory, a property which survives even though there is a spontaneous symmetry breakdown which generates the masses of the weak bosons.

Over the last two years, there has been dramatic advance in this field of physics, with the successes of the CERN proton–antiproton Collider at Geneva. This has placed us abruptly in a much higher energy regime where the weak and electromagnetic interactions have comparable strength, enabling the observation and measurement of the $W^\pm$ and Z^0 bosons, and giving a most direct verification of the standard model. Also in these experiments, one observes the scattering of quarks and gluons giving rise to jets of mesons and baryons due to the hadronisation of an outgoing quark or gluon. These are rare processes, but most informative, telling us, for example, that quarks and gluons have no structure, at least down to a radius of order 10^{-18} m. The energy available should be ample for the production of the top quark, unless it is quite unexpectedly heavy, and it is hoped that future Collider experiments may find evidence for this particle.

There is another regime of importance in particle physics at high energies, not much emphasised today, the regime of 'soft' processes at high energies but small energy–momentum transfer. This is the regime for which the Regge representation for scattering and reaction amplitudes was devised. This representation is not so commonly discussed these days but it still provides an essential basis for the analysis of a wide range of soft hadronic processes. A central element of the Regge representation is the Regge trajectory, which links hadronic spectroscopy in the region $q^2 = m^2 > 0$ with the differential cross sections for the low-momentum-transfer exclusive processes induced by exchange of the corresponding Reggeon in the region $q^2 < 0$. With the high-energy meson beams available with the SPS accelerator at CERN and with the proton accelerator at FNAL, some of these trajectories, the ρ- and A_2-trajectories in particular, have been traced experimentally quite far into the regime $q^2 < 0$.

However, there is one Reggeon, if Reggeon it be, which has quite different behaviour. In the Regge picture, hadron–hadron elastic scattering at small angles and total cross sections are dominated by Pomeron exchange, which, in the simplest view, would require that all hadronic total cross sections approach constant values at infinite energy. In fact, the cross sections are observed to be rising quite strongly with increasing energy, on a logarithmic scale, and the cosmic ray data suggest that they continue to rise at still higher energies. It is not excluded that the cross sections may level off with increasing energy, ultimately reaching constant limiting values; the question is how high the energy must be for the asymptotic expressions to hold valid. Viewed from the standpoint of QCD, the Pomeron results from a special subset of gluonic exchanges, but it is far from obvious that these dominant graphs actually correspond to a single Pomeron exchange. The multiple exchange of Pomerons will also contribute, and their net effect is somewhat uncertain. Different people considering this question reach rather different conclusions. Thus, there is still doubt about the nature and status of the Pomeron singularity and its trajectory, and one should feel some reserve about its unqualified use.

In 1981, Peter Collins and Alan Martin wrote an important report (*Rep. Prog. Phys.* 1982 **45** 335) endeavouring to present a coherent view of all of hadronic physics, especially concerning the high-energy regime as then known from SPS and ISR experiments at CERN and from FNAL experiments in America, within the general framework of quantum chromodynamics with quarks. This report stressed that, with quantum chromodynamics, the Regge representation is still relevant and useful, and emphasised that a unified view of data coming from quite different directions is possible within this framework, and that much of the formalism developed in an earlier era is still applicable. We urged the authors to consider extending their useful report to book length, bringing it up-to-date to the end of 1983. This has proved to be a very appropriate time in the development of the subject, as most of the analyses of the first major data runs at the CERN proton–antiproton Collider have recently been completed. The most striking conclusions have been published, or presented as conference reports. They have turned out to be quite compatible with the data at lower energies which all fit together within the same picture. The CERN Collider is now out of commission for more than a year and we cannot hope to see much new data before 1985. It is therefore rather appropriate to publish now a book which surveys all of the physics of high-energy interaction phenomena, from hadron spectroscopy to jet production in high-energy proton–antiproton collisions, within a single framework, that of the symmetry $SU(3)_C \times SU(2) \times U(1)$.

Because of the coherence of the data, this book is based on the authors' article in *Reports on Progress in Physics* but is extended to include the experimental domain of the CERN Collider. The earlier sections concerning energies up to that of the ISR have been revised and corrected wherever necessary. However, surprisingly few changes were needed. The new information has fitted in with reasonable extrapolations from the lower energies in a satisfactory way. Jet phenomena are now very clear in the Collider experiments, whereas they were uncertain at ISR energies.

This volume discusses high-energy particle physics up to the end of 1983, and the concepts it brings together and the overall picture it gives appear likely to dominate until late in this decade. The information is presented in a solid theoretical framework, that of QCD. No doubt new phenomena may be found even within this period, but these are likely still to lie within the general framework used as the basis of this book. It should thus provide a good starting point for approaching these new developments. We hope that it will prove useful to young physicists coming into this field, as well as to experienced physicists transferring into this area of very high-energy interactions.

R H Dalitz   Oxford
                                           13 January 1984

CONTENTS

ACKNOWLEDGMENTS

It is a pleasure to thank our colleagues who have helped in the preparation of this book. We are especially grateful to Arthur Clegg who asked us to write the original article for *Reports on Progress in Physics* 1982 **45** 335, and to Dick Dalitz who not only encouraged us to extend it into a book but who has carefully read the entire manuscript and suggested many improvements. We also thank Peter Watkins for providing us with diagrams of events obtained by the UA1 detector at CERN, and Mike Albrow, Francis Halzen, Maurice Jacob, Mike Pennington, John Rushbrooke and David Scott for critical reading of, and valuable comments on, the later chapters. We are also grateful to Anthony Allan, Nigel Glover, Neil Speirs and Tim Spiller for their comments, and to Vicky Kerr and Muriel Raines for their efficient typing of the manuscript.

P D B Collins
A D Martin

1

INTRODUCTION

1.1 Partons

The last decade has seen the discovery of a completely new branch of physics, the physics of the constituents of 'elementary' particles, which are usually called 'partons'. These partons come in two kinds; the quarks which have spin $\frac{1}{2}\hbar$ and carry such properties as charge, isospin, strangeness, etc, and the gluons which have spin $\hbar$. Quarks and gluons interact with each other because they carry a new sort of charge called 'colour'. The elementary particles of nuclear physics with which we have been familiar for much longer, like the proton, neutron, pi meson or Λ hyperon for example, are now regarded as composite particles constructed from these quarks and gluons.

This revelation of a new layer of the sub-structure of matter is, of course, in many respects a repetition of a familiar pattern in the history of physics. The 19th century saw the establishment of atoms as the basic elementary components from which chemical compounds (molecules) are constructed. Then came the discovery of the electron and the nucleus, and in the first 30 years of this century the quantum theory of the structure of atoms was developed. They are quantum bound states of negatively charged electrons and a positive nucleus held together by electromagnetic forces due to the exchange of virtual photons. During the 1930s and 1940s a theory of atomic nuclei, as composites of varying numbers of protons and neutrons bound by the nuclear (or strong) interaction force stemming from the exchange of mesons, was constructed.

However, in the 1950s and 1960s experiments involving the scattering of these supposedly elementary particles at high energies produced very large numbers of other, similar, strongly interacting particles, and physicists were forced to recognise that protons and neutrons are simply the lightest members of a very large family of particles called 'baryons'. About 90 different species of baryon have been identified to date (Particle Data Group 1982), but most are highly unstable and decay after very short times (about 10^{-23} s) which is why they are not so well known. Similarly, since the postulation by Yukawa in 1935 of the 'meson' as the carrier of the nuclear force, and discovery of the pi meson in 1947, more than 70 additional mesons have been found (Particle Data Group 1982), and there is no reason to believe that their number has been exhausted. This very large group of particles which enjoy the strong interaction, the baryons and the mesons, are called 'hadrons' (from the Greek, meaning 'large').

Already in the early 1960s the suggestion had been made that hadrons could conveniently be regarded as composites of more basic objects called 'quarks' which carry charge, isospin and strangeness (Gell-Mann 1964, Zweig 1964). At that time many physicists regarded these quarks as little more than a convenient mathematical device with which to model the properties of hadrons, since after all free quarks were not seen. However, it was not too surprising that when experiments to probe the structure of the proton by scattering electrons began at Stanford in the late 1960s it was revealed that protons did indeed contain point-like constituents. These were called 'parts' or 'partons' by Feynman (1969) and are now recognised as being the quarks of the original quark model. During the 1970s colliding-beam experiments of the type $e^+e^- \to$ hadrons have revealed the existence of further types of quarks carrying new properties such as charm and beauty, so we now know of quite a number of different partons (see table 1).

Table 1. The flavour quantum numbers of quarks and leptons. Q, S, C, B, T are charge, strangeness, charm, beauty and truth. Free quarks are not seen and the mass represents the current quark mass felt via electromagnetic or weak interactions. All quarks and leptons have antiparticles (with opposite $Q, S, \ldots$). Each kind, or flavour, of quark comes in three colours (red, green, blue). The three different generations are indicated.

	Quarks						Leptons			
Flavour	Mass (GeV/c^2)	Q	S	C	B	T		Mass (GeV/c^2)	Q	Generation
Down, d	0.008	$-\frac{1}{3}$	0	0	0	0	e^-	0.0005	-1	⎫ 1
Up, u	0.004	$+\frac{2}{3}$	0	0	0	0	ν_e	0	0	⎭
Strange, s	0.15	$-\frac{1}{3}$	-1	0	0	0	μ^-	0.105	-1	⎫ 2
Charm, c	1.2	$+\frac{2}{3}$	0	1	0	0	ν_μ	0	0	⎭
Bottom, b	4.7	$-\frac{1}{3}$	0	0	-1	0	τ^-	1.8	-1	⎫ 3
Top, t?	?	$+\frac{2}{3}$	0	0	0	1	ν_τ	0	0	⎭

This probing of the constitution of matter has been made possible by the development of accelerators of increased energy. A particle beam of momentum p has an associated wavelength $\lambda = h/p$ which, according to Heisenberg's uncertainty principle, determines the best spatial resolution which that beam can provide. For numerical estimates it is useful to recall that $\hbar c \equiv hc/2\pi \approx 2 \times 10^{-7}$ eV m $= 0.2$ GeV fm. Thus, to determine the electron distribution in an atom we are concerned with distances of the order of 1 Å $= 10^{-10}$ m and hence need electron beams with energies about 10^3 eV, while to probe the charge distribution of a nucleus at distances of a few fm needs 0.1 GeV electrons, but to observe the parton structure of a proton at distances very much less than 1 fm requires electrons with at least 10 GeV energy.

Electron scattering experiments, and more recently muon scattering too, probe the charge structure of hadronic matter. Complementary information is obtained with neutrino beams which couple to matter via the weak interaction only, and hence probe its weak structure, while the scattering of hadrons by hadrons enables us to explore the strong interaction structure. The parton concept helps us to understand all three types of experiment. Our intention in this review is to try to illustrate how the parton sub-structure of hadrons determines the results of such experiments, but first we must introduce the various types of particles with which we shall be dealing, and the interactions which they undergo.

1.2 Leptons and quarks

Apart from gravity, which we shall not need to consider in this review, particles undergo three seemingly quite different types of interaction; the electromagnetic interaction of charged particles, the short-range weak interaction which is responsible for the β decays of nuclei for example, and the strong or hadronic force which, *inter alia*, binds nucleons into nuclei.

There is one group of particles, spin $\frac{1}{2}\hbar(\equiv \mathrm{spin}\ \frac{1}{2})$ fermions called leptons, which do not experience the hadronic force but only electromagnetism and the weak interaction (see table 1). The best known of these is the electron, e^-, and with it are associated its antiparticle the positron, e^+, and the electron's neutrino and antineutrino, $\nu_e, \bar\nu_e$, which are produced along with the electron (or positron) in β decay (for example, $n \to p + e^- + \bar\nu_e$). A second family of leptons consists of the muon, μ^-, together with $\mu^+, \nu_\mu, \bar\nu_\mu$ (ν_μ being a different kind of neutrino). The μ appears to be identical to e in most respects except that it is about 200 times heavier, and is unstable, decaying by $\mu^- \to e^- + \bar\nu_e + \nu_\mu$, with a lifetime $\approx 10^{-6}$ s. More recently the τ lepton family has been discovered (Perl et al 1975) in the experiment $e^+e^- \to \tau^+\tau^-$ and although the existence of its neutrino ν_τ is not quite firmly established the expected decays $\tau^- \to e^- + \bar\nu_e + \nu_\tau$ and $\tau^- \to \mu^- + \bar\nu_\mu + \nu_\tau$ do seem to occur (as well as $\tau^- \to$ hadrons $+ \nu_\tau$). No further charged leptons are known, up to a mass of $20\ \mathrm{GeV}/c^2$ at least (Bartel et al 1983).

All these leptons seem to be structureless points, i.e. they have no apparent size. This is deduced from scattering experiments such as $e^+e^- \to e^+e^-$ and $e^+e^- \to \mu^+\mu^-$ which attempt to probe their structure, but reveal none down to about 10^{-2} fm (Bohm 1980), but a more precise (if theoretically controversial) limit on the size of charged leptons ($<10^{-6}$ fm) can be deduced from the fact that their anomalous magnetic moments ($g - 2$) are in accord with the predictions of quantum electrodynamics (QED), to a few parts in 10^{-10} for the electron (Brodsky and Drell 1980). If the charge distribution occupied a finite volume this would obviously change the gyromagnetic ratio g from the Dirac value of 2. We thus regard the leptons as elementary (so far (!), but see Harari (1980) for a discussion of possibilities for composite leptons).

Hadrons, such as the proton, are not elementary however. The proton's gyromagnetic ratio is 5.56, not 2, indicating that its charge is distributed over a finite volume ($\approx 1\ \mathrm{fm}^3$). This is because a proton is made up of more fundamental charged constituents, the quarks (see, for example, Close (1979) for a review of quark models).

Quarks are also spin-$\frac{1}{2}$ fermions with fractional electric charges, as shown in table 1, and are also point-like to the accuracy of present electron-scattering data (see chapter 3). In addition to the weak and electromagnetic interactions, quarks experience the strong colour force which, it is believed, is so strong as to confine them within hadrons. This is presumably why free quarks have not been found (see the next section). Certainly quarks cannot be knocked out of hadrons by any of the high-energy beams available at present, and the only positive evidence for free quarks is extremely controversial. Despite this the quarks are easily 'seen' in deep inelastic e, μ and ν scattering experiments on protons. But since the quarks cannot be taken out and weighed the evidence for their masses quoted in table 1 is somewhat indirect (see Gasser and Leutwyler 1982).

The quark composition of some of the lighter mesons ($q\bar q$) and baryons (qqq) is indicated in table 2. This (constituent) quark model has been remarkably successful in hadron spectroscopy (see, for example, Hey and Morgan 1978, Close 1979). Multiplets of the predicted spins and parities are found, and many members of the multiplets clearly identified. No states outside the quark model predictions, or multiquark states containing more than the minimal number of quarks (such as $qqqq\bar q$), have been established experimentally. The quarks shown in table 2 are the so-called 'valence' quarks which are needed to carry the quantum numbers (charge, strangeness, charm, etc) which are often referred to as 'flavours'.

Table 2. The quark content of the lighter meson and baryon multiplets. The particle masses are listed in GeV/c^2. The hadrons made of n types (flavours) of quarks can be grouped into multiplets of SU(n). For the mesons ($q\bar{q}$) we show the members of the $15+1$ dimensional multiplets of SU(4) of spin, parity $J^P = 0^-$, 1^-, 2^+; while for the baryons (qqq) we just show the $\frac{1}{2}^+$ octet and $\frac{3}{2}^+$ decuplet of SU(3), although some charmed baryons have been identified. For u, d, s composites this is the famous SU(3) symmetry of Gell-Mann (1961) and Ne'eman (1961). The mass spread within a multiplet is indicative of symmetry breaking.

(*a*) Meson multiplets

	0^-	Mass	1^-	Mass	2^+	Mass
$u\bar{d}$, $d\bar{u}$	$\pi^\pm$	140	$\rho^\pm$		$A_2^\pm$	
$(u\bar{u} - d\bar{d})/\sqrt{2}$	π^0	135	ρ^0	770	A_2^0	1317
$u s$, $s u$	$K^\pm$	494	$K^{*\pm}$	892	$K^{*\pm}$	
$d\bar{s}$, $s\bar{d}$	K^0, $\bar{K}^0$	498	K^{*0}, $\bar{K}^{*0}$	898	K^{*0}, $\bar{K}^{*0}$	1434
$(u\bar{u} + d\bar{d})/\sqrt{2}$	η	549	ω	782	f	1273
$s\bar{s}$	η'	958	φ	1020	f'	1516
$c\bar{d}$, $d\bar{c}$	$D^\pm$	1868	$D^{*\pm}$	2009	$D^{*\pm}$	?
$c\bar{u}$, $u\bar{c}$	D^0, $\bar{D}^0$	1863	D^{*0}, $\bar{D}^{*0}$	2006	D^{*0}, $\bar{D}^{*0}$	?
$c\bar{s}$, $s\bar{c}$	$F^\pm$	1970	$F^{*\pm}$	2140?	$F^{*\pm}$	?
$c\bar{c}$	η_c	2980	ψ	3097	χ	3551

(*b*) Baryon multiplets

	$\frac{1}{2}^+$	Mass	$\frac{3}{2}^+$	Mass
uuu, ddd			Δ^{++}, Δ^-	1232
uud, udd	p, n	939	Δ^+, Δ^0	
uus, uds, dds	$\Sigma^{+,0,-}$	1195	$\Sigma^{*+,0,-}$	1385
uds	Λ	1116		
uss, dss	Ξ^0, Ξ^-	1318	Ξ^{*0}, Ξ^{*-}	1533
sss			Ω^-	1672

There is one feature of table 2 which used to seem particularly puzzling. It will be noted that the Δ^{++}, for example, consists of three u quarks, all with their spins parallel to make a spin-$\frac{3}{2}$ state, with no orbital angular momentum. But quarks are fermions, and by the exclusion principle it should not be possible to have three identical quarks in the same state. (Compare the Li atom ground state where the two 1s electrons have opposite spin orientations while the third has $l = 1$ giving an overall $L = 1$, $S = \frac{1}{2}$ state.) The now generally accepted explanation for this apparent anomaly is that quarks possess an additional property, called 'colour', which can take three possible values, say red, green and blue (see, for example, Greenberg and Nelson 1977).

The hadrons are postulated to be colourless (white), i.e. they contain equal mixtures of red (R), green (G) and blue (B) quarks. Thus the Δ^{++} state is $u^R u^G u^B$ and so no longer contains identical fermions. More generally we write for a meson with the valence quark structure $q_1\bar{q}_2$ (table 2)

$$M = \frac{1}{\sqrt{3}}(q_1^R\bar{q}_2^R + q_1^B\bar{q}_2^B + q_1^G\bar{q}_2^G) \tag{1.1}$$

while a $q_1 q_2 q_3$ baryon becomes

$$B = \frac{1}{\sqrt{6}} \epsilon_{\alpha\beta\gamma} q_1^\alpha q_2^\beta q_3^\gamma \qquad \text{with } \alpha, \beta, \gamma = \text{R, G, B} \qquad (1.2)$$

(where $\epsilon_{\alpha\beta\gamma}$ is the antisymmetric permutation tensor) ensuring overall antisymmetry.

The quarks are regarded as the fundamental triplets of an $SU(3)_C$ colour gauge symmetry group (Abers and Lee 1973, Politzer 1974, Iliopoulos 1976). There is now abundant evidence for the existence of the three colour degrees of freedom for each flavour of quark, so the quarks appearing in tables 1 and 2 should really have colour labels R, G, B appended as above. The leptons, on the other hand, are colourless and hence not bound into hadrons.

It will be noted that we have associated quarks and leptons together to form different 'generations' in table 1. An extensive discussion of the justification for this will be found in Harari (1978). Quite apart from the obvious similarity of quarks and leptons, as point-like fermions having electromagnetic and weak interactions (though only the former participate in the strong interaction), there are theoretical reasons for believing that in order to ensure that a proper renormalisable field theory can be constructed the sum of the charges of all the fundamental fermions must vanish. It will be seen from table 1 that

$$Q_e + Q_\nu + 3(Q_u + Q_d) = 0 \qquad (1.3)$$

(the factor of 3 stemming from the three colours of quarks) so the relation is satisfied by each generation of fermions separately.

1.3 The strong interaction

The most familiar type of force experienced by these particles is the electromagnetic interaction, which is due to the exchange of massless vector (i.e. spin 1) virtual photons (the quanta of the electromagnetic field) between charged particles (Bjorken and Drell 1964). The lowest-order diagram of QED involves the exchange of a single photon between an electron and a positron, as shown in figure 1(a). It gives rise to the Coulomb interaction potential

$$V(r) = -\alpha/r \qquad (1.4)$$

where $\alpha \equiv e^2/4\pi\hbar c \simeq 1/137$ is the fine-structure constant. Higher-order contributions such as figure 1(b) involve more couplings and hence are smaller by further powers of α, and so can often be neglected. However, it is a very important matter of principle

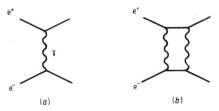

Figure 1. $O(\alpha)$ and $O(\alpha^2)$ photon exchange contributions to the e^+e^- QED interaction.

that the physical coupling increases with $Q^2 \equiv -q^2 > 0$ where q is the four-momentum of the virtual photon. This is due to vacuum polarisation effects that shield the bare charge. Including these diagrams, figure 2, gives the leading behaviour

$$\alpha(Q^2) \simeq \alpha \left[1 + \frac{\alpha}{3\pi} \log \left(\frac{Q^2}{\mu^2} \right) + \left(\frac{\alpha}{3\pi} \log \left(\frac{Q^2}{\mu^2} \right) \right)^2 + \ldots \right] \tag{1.5}$$

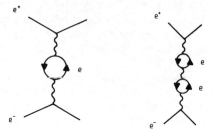

Figure 2. Vacuum polarisation corrections to the electric charge, see figure 1(a).

where μ is the arbitrary normalisation point at which α has been measured. In higher orders a whole series of $[\log (Q^2/\mu^2)]^n$ terms appears (for a simple account see Close 1982). These leading logs may be summed with the result

$$\alpha(Q^2) \simeq \frac{\alpha(\mu^2)}{1 - [\alpha(\mu^2)/3\pi] \log (Q^2/\mu^2)} \tag{1.6}$$

for $Q^2 \gg \mu^2$. We thus have a 'running' coupling constant, not really constant at all, which changes (runs) with Q^2. As Q^2 increases the photon sees more and more of the bare charge, and at some very large, but finite, Q^2 the coupling $\alpha(Q^2)$ is infinite. The bare charge is thus ultraviolet-divergent, but this does not usually worry us since we expect that quantum gravity effects will have modified the theory long before such very large Q^2 (very short distances) are reached. In the infrared region of experimental QED the Q^2 dependence of α is essentially undetectable; a 10% effect requires a Q^2 range of $\exp(0.3\pi/\alpha) \simeq 10^{56}$. For all practically attainable Q^2 we can expect perturbation theory to be satisfactory.

The strong interaction is the result of the exchange of coloured massless vector gluons between coloured quarks (see Marciano and Pagels 1978). We noted in § 1.2 that three colours of quark are necessary (R, G, B) and that these are regarded as the fundamental triplet of an $SU(3)_C$ coloured gauge symmetry group. The emission of a gluon can change the colour of a quark (see figure 3(a)) and so there appears to be nine possible colours of gluons (R$\bar{\text{R}}$, R$\bar{\text{G}}$, R$\bar{\text{B}}$, G$\bar{\text{G}}$, G$\bar{\text{R}}$, G$\bar{\text{B}}$, B$\bar{\text{B}}$, B$\bar{\text{R}}$, B$\bar{\text{G}}$). However, the combination $(1/\sqrt{3})(\text{R}\bar{\text{R}} + \text{B}\bar{\text{B}} + \text{G}\bar{\text{G}})$ is colourless and so does not couple to the quarks (and hence if it existed would not be detectable). We thus have an octet of vector gluons in the adjoint representation of the $SU(3)_C$ group. The quark–gluon vertex coupling (figure 3(b)) may then be expressed as $\frac{1}{2}g_s\lambda_{ij}^\alpha$ where g_s is the strong coupling (analogous to the charge e of QED) and the λ_{ij}^α are Gell-Mann's $SU(3)$ representation matrices. There are eight of them, with $\alpha = 1, \ldots, 8$ for the gluon colours, while $i, j = 1, 2, 3$ for the quark colours.

Superficially this quantum chromodynamics (QCD) description of the strong interaction, with massless vector bosons exchanged between fermions, seems very similar to

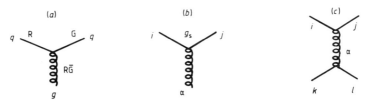

Figure 3. (a), (b) The quark–gluon vertex, and (c) the lowest-order $q\bar{q}$ interaction with colour labels.

QED in that the lowest-order diagram, figure 3(c), involving a single gluon exchanged between a quark and an antiquark gives

$$V_{ijkl}(r) = - \sum_{\alpha=1}^{8} \tfrac{1}{4} \lambda_{ij}^{\alpha} \lambda_{lk}^{\alpha} (\alpha_s / r) \tag{1.7}$$

where $\alpha_s \equiv g_s^2 / 4\pi$. For colourless initial and final states (when the q and $\bar{q}$ form a meson, for example), since

$$\sum_{\alpha} \lambda_{ij}^{\alpha} \lambda_{ji}^{\alpha} = 16/3 \tag{1.8}$$

we find

$$V(r) = -\tfrac{4}{3}(\alpha_s / r) \tag{1.9}$$

just like (1.4) except for the colour factor $\tfrac{4}{3}$. However, the effect of higher-order diagrams like figure 4 gives

$$\alpha_s(Q^2) \simeq \alpha_s \left\{ 1 - \frac{\alpha_s b_0}{4\pi} \log\left(\frac{Q^2}{\mu^2}\right) + \left[\frac{\alpha_s b_0}{4\pi} \log\left(\frac{Q^2}{\mu^2}\right)\right]^2 + \ldots \right\} \tag{1.10}$$

$$\simeq \frac{\alpha_s}{1 + (\alpha_s b_0 / 4\pi) \log\,(Q^2/\mu^2)} \equiv \frac{1}{(b_0/4\pi) \log\,(Q^2/\Lambda^2)}$$

$$\Lambda^2 \equiv \mu^2 \exp\,(-4\pi/\alpha_s b_0) \tag{1.11}$$

where μ^2 is the value of Q^2 at which α_s is measured, and $b_0 \equiv \tfrac{11}{3} N_c - \tfrac{2}{3} N_f$ where N_c is the number of colours ($= 3$) and N_f is the number of flavours of quark ($= 6$ in table 1). The N_c term stems from the gluon loop (figure 4(b)) which arises because gluons carry colour and hence couple to each other. There is no similar diagram in QED (figure 2) because photons do not carry the charge coupling. The N_f term arises from the quark loop (figure 4(a)) and is just the same as figure 2 of QED except that, of course, we must sum over all the flavours of quarks which can contribute.

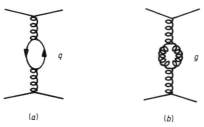

Figure 4. The lowest-order corrections to the quark–gluon coupling.

Thus, with $N_c = 3$ and $N_f = 6$ we have $b_0 = 7$, and as long as $N_f < 16$, we always find $b_0 > 0$ so that the sign in the denominator of (1.11) is opposite to that in (1.6). This has the important consequence that $\alpha_s(Q^2) \to 0$ as $Q^2 \to \infty$ which means that quarks and gluons appear like almost-free particles when looked at with very high-energy probes, which are sensitive to the short-distance structure of the hadron. This 'asymptotic freedom' (Politzer 1973, Gross and Wilczek 1973) will be an essential ingredient of the parton approach to the structure of hadrons which we shall explore in this book.

The other important consequence of (1.11) is that $\alpha_s(Q^2) \to \infty$ as $Q^2 \to \Lambda^2$ (which serves to define Λ) and so the perturbation series breaks down at small Q^2. Taking the Fourier transform of (1.11) we obtain

$$\alpha_s(r) \simeq \frac{1}{(b_0/2\pi) \log (1/\Lambda r)} \qquad (1.12)$$

and so the coupling becomes stronger as the separation between the q and $\bar{q}$ increases, and the perturbation series breaks down as $r \to \Lambda^{-1}$. This is because of the gluon self-coupling, which implies that the exchanged gluons will attract each other (unlike photons) and so the colour lines of force are constrained to a tube-like region between the quarks (unlike the Coulomb field in which the lines of force just spread out) (see figure 5). If these tubes have a constant energy density per unit length then the potential energy of the interaction will increase with the separation, $V(r) \sim \lambda r$, and so the quarks (and gluons) can never escape from the hadron. This so-called 'infrared slavery' is believed to be the origin of the confinement mechanism and explains why we do not observe free quarks (Feynman 1972, Dokshitzer $et\ al$ 1980). They have very little chance of straying outside the confinement range $\sim \hbar c/\Lambda$. So the fact that hadrons have a size ~ 1 fm suggests that $\Lambda \simeq 0.2$ GeV.

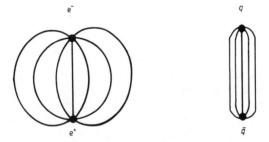

Figure 5. The $e^+ e^-$ Coulomb field with potential $V(r) \sim 1/r$, and the $q\bar{q}$ colour field with $V(r) \sim r$.

Unfortunately, practical calculations in field theory generally invoke perturbation methods which are clearly not applicable to the bound-state problem of QCD with $\alpha_s \gtrsim 1$, and so far it has not proved possible to demonstrate conclusively that confinement is a consequence of QCD. However, most physicists are now fairly confident that this is the right approach, partly because of the remarkable success of essentially non-relativistic models of hadrons based on these ideas.

Thus one might guess that the effective interaction potential between q and $\bar{q}$ in a meson could be approximated by a combination of a short-range asymptotic freedom contribution due to single-gluon exchange and a long-range confining potential which increases with r, such as

$$V(r) = -\tfrac{4}{3}\frac{\alpha_s}{r} + \lambda r. \tag{1.13}$$

This sort of potential, substituted into the Schrödinger equation, gives a very good account of the spectrum of mesons made of heavy quarks such as $\psi(c\bar{c})$ and its excited (charmonium) states, and of the $\Upsilon(b\bar{b})$ spectrum (Quigg and Rosner 1979, Isgur and Karl 1979, Eichten *et al* 1980). Indeed, if the usual non-relativistic reduction of the one-gluon exchange term is used to generate hyperfine spin–spin and spin–orbit interactions, just like the Fermi–Breit Hamiltonian of atomic physics but with 'colour magnetic moments', many detailed features of both the meson and baryon spectrum can be explained with very few arbitrary parameters. The only unknowns are the quark masses and the couplings α_s, λ. We can thus take quite seriously the idea that hadrons are made of quarks bound together by a confining potential due to the exchange of gluons.

But this sort of approach only has theoretical plausibility for states which are made of heavy quarks like c and b, so that the binding energy is very small compared to the quark mass, and hence the internal motions are not too relativistic ($v/c \ll 1$). For particles which are composed entirely of u, d and s quarks, whose masses are $\leqslant \Lambda$, the QCD scale, the coupling α_s is $\geqslant 1$ and so the probability of creating additional virtual gluons and $q\bar{q}$ pairs within the hadron becomes very great. In these circumstances it seems better to regard the hadron as made up of three classes of constituent (Feynman 1972, Llewellyn-Smith 1972, Kogut and Susskind 1974):

(i) the 'valence' quarks q_v which carry its quantum numbers such as charge, strangeness, etc (for example, the proton consists of $u_v u_v d_v$ valence quarks according to table 2);

(ii) a 'sea' of virtual gluons, the quanta of the colour force field which are exchanged between the quarks, and between the gluons themselves;

(iii) a 'sea' of quarks and antiquarks, i.e. virtual $q_s\bar{q}_s$ pairs created by vacuum polarisation of the colour field as in figure 4.

Whereas the number of valence quarks is fixed by the quantum numbers of the particle, the number of virtual sea quarks and gluons is unlimited and rapidly fluctuating. As the gluons are massless there is no inhibition to the creation of large numbers of low-energy gluons and so the probability of finding such gluons can be expected to diverge as their energy tends to zero (the infrared divergence). One also expects large numbers of u and d sea quarks because they are very light compared to Λ (see table 1) but fewer s quarks, while the number of heavy c or b quarks is presumably quite small because of the high-energy 'cost' of creating them (virtually). More accurately, the uncertainty principle requires that a virtual state of mass M can exist only for times Δt such that $\Delta t < \hbar/Mc^2$ so high-mass $q\bar{q}$ pairs can only be present for a very small fraction of the time.

It is this essentially many-body nature of all relativistic bound-state systems which renders the perturbation methods of quantum field theory impotent. It might therefore seem that there is no hope of gaining any understanding of such complex structures, but fortunately this is not so. We can, in fact, obtain a good deal of experimental information about the structure of hadrons by probing them in various types of scattering experiment, and many features of the results can be explained through the parton model.

1.4 Deep inelastic scattering

Even though confinement prevents us from taking a quark out of a hadron for detailed examination, it is still possible to 'see' quarks, for example in lepton scattering experiments (Gilman 1972, Feynman 1972). The situation is perhaps somewhat analogous to the model 'ship in a bottle' which sailors used to make. One can readily see the ship because of the light it scatters, but one cannot get it out. If one wants to obtain a really detailed description of the ship one may need to make allowance for the refractive properties of the glass bottle which distorts its image—i.e. one needs to have an understanding of the confining mechanism.

One way of seeing the quarks in a hadron is by electron-scattering experiments. Electrons have been used for many years to probe the structures of atoms. As noted in § 1.1 one needs electrons of a few keV energy to resolve to better than 1 Å. A typical result is shown in figure 6(a) which gives the cross section as a function of the energy of the electron after scattering. There is a large elastic peak in which the electron has been scattered by the nucleus (with negligible loss of energy because the nucleus is so much heavier and barely recoils) and then a smaller 'quasi-elastic' peak in which the electron has hit one of the electrons in the atom, and lost on average half its energy in the process. However, because the atomic electron is confined to move in orbit round the nucleus, this peak has a width determined by the dispersion of the momentum of the bound electron, and hence, through the uncertainty principle, by the size of the atom.

To discover the structure of the nucleus one needs to scatter electrons having energies of a few hundred MeV, to give a resolution of the order of a fermi; see figure 6(b). There is an elastic peak due to scattering off the nucleus as a whole, and a

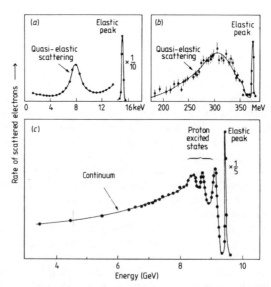

Figure 6. Electron scattering data, in different energy regimes, compiled by Amaldi (1979). (a) e⁻ + carbon, $E = 15$ keV, $\theta = 45°$; (b) e⁻ + helium, $E = 400$ MeV, $\theta = 60°$; (c) e⁻ + proton, $E = 10$ GeV, $\theta = 10°$, showing the energy of the scattered electron.

quasi-elastic peak, in which the electron has been scattered by an individual proton, whose width indicates the dispersion of the proton's Fermi momentum.

Similarly one can probe the structure of an individual proton by scattering electrons of a few GeV energy. Again there is an elastic peak in which the proton recoils as a whole, some subsidiary peaks due to the excitation of the proton to various higher-mass N* resonant states, and then a continuum distribution of those electrons which have been scattered by the constituents of the proton, i.e. the quarks. This is the so-called 'deep inelastic scattering' (DIS). As we shall see in chapter 3 the size of this continuum cross section, and its variation with energy, scattering angle, etc, is just what one would expect if the scattering were due to free, point-like, spin-$\frac{1}{2}$, charged particles. The quarks appear to be free (i.e. there is rather little effect of the strong force which binds the quarks into the hadron) because we are using a large momentum probe, and $\alpha_s(Q^2) \to 0$ as $Q^2 \to \infty$. The success of the parton model depends crucially on this asymptotic freedom.

Similar experiments can be carried out using neutrino beams (Llewellyn-Smith 1972) instead of electrons, probing the weak interaction structure rather than the charge distribution, and entirely consistent results are obtained (see chapter 4). The gluons have no electromagnetic or weak interaction and so are not seen directly in these DIS experiments, but their properties can be inferred indirectly, as we shall see.

But what happens to the quarks after they have been struck hard by the electron or neutrino? In the corresponding atomic (or nuclear) scattering experiments the constituent electrons (or protons) are knocked out of the composite system and so can be detected as free particles in the final state. However, we do not obtain free quarks as a result of DIS. The reason for this is, of course, confinement, which we believe operates something like figure 7 (Bjorken 1973, Dokshitzer et al 1980).

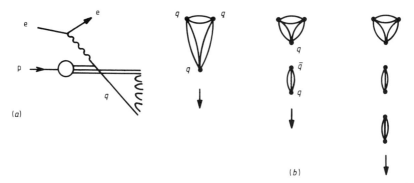

Figure 7. Deep inelastic electron–proton scattering. Diagram (b) shows, at successive time intervals, the struck quark leaving the proton. The hadronisation, or creation of $q\bar{q}$ pairs, which is sketched in diagram (a) in the form of a chain, neutralises the colour in the final state.

The struck quark in figure 7(a) attempts to leave the proton, but in so doing it stretches the colour lines of force into a tube (cf figure 5) until the potential energy of the colour field is sufficient to create a $q\bar{q}$ pair. These can act as the end points for the lines of force, which thus break into two shorter tubes with lower net energy (despite the extra $q\bar{q}$ mass) than the single extended tube (figure 7(b)). The outgoing quark continues on its way, stretching the lines of force, and further $q\bar{q}$ pairs are produced, until eventually all its kinetic energy has been degraded into clusters of quarks and gluons, each of which has zero net colour and low internal momentum. These clusters can (indeed must) form hadrons since now $\alpha_s(Q^2) \gtrsim 1$, and so the energy given to the struck quark finally manifests itself as a 'jet' of hadrons travelling more or less in the direction of that quark (Konishi *et al* 1979). The quark thus escapes from its parent hadron, but only into another, newly created, faster moving hadron. The price of asymptotic freedom is infrared slavery!

Hence DIS is followed by a process of 'hadronisation' in which the energy lost by the electron during the scattering process is converted, via colour polarisation of the vacuum, into new hadrons. We shall be examining these jet phenomena in more detail in chapter 5.

1.5 Exchange forces

We have seen that fundamentally the strong interaction seems very similar to QED in that the basic interaction, massless vector gluon exchange between coloured quarks, is very like massless vector photon exchange between charged electrons. The crucial difference is that gluons carry colour and hence couple to each other, whereas photons do not carry charge. This means that the strong coupling constant 'runs' in the opposite direction to the electromagnetic coupling, and becomes very large at low momenta, ensuring that the coloured quarks and gluons are confined within colourless hadrons. Any attempt to knock them out simply produces more hadrons.

The strong interaction which we observe most directly, therefore, is not that between the quarks themselves, but between the composite hadrons. The nuclear force which binds protons and neutrons into nuclei, for example, is the residual colour polarisation force between these colourless hadrons. This is analogous to molecular binding forces, which can be regarded as the residual electromagnetic polarisation effects between neutral composite atoms. Its principal manifestation is an exchange force, involving the exchange of electrons between the atoms. Similarly the nuclear force involves the exchange of coloured quarks and gluons between the hadrons.

By the uncertainty principle the range of an exchanged particle of mass m is given by $r = \hbar/mc$ and so the longest-range part of the p–n force is provided by the exchange of the lightest colourless composite object which can be made up from quarks and gluons. From table 2 we see that this is the pion. This is just what Yukawa suggested in 1935, except that we now regard it as the exchange of a $q_v\bar{q}_v$ valence pair with attendant sea quarks and gluons (figure 8(a)), not of an elementary particle.

But of course the $q\bar{q}$ pair need not be in a pion, but in any of the other more massive meson states which can be made with these quarks (see table 2) such as the ρ or A_2. These heavier states provide shorter-range forces of course. This partly explains why the short-range part of the nuclear force is so complicated—there are so many massive mesons which can be exchanged. In fact, if we are interested in high-energy hadron scattering we are forced to take into account the exchange of all these particles together. A way of doing this was discovered (in a quite different context) by Regge in 1959 (see Collins 1977).

If one regards the mesons as $q\bar{q}$ bound states produced by an effective gluon exchange potential $V(r)$ like (1.13) (see figure 5), then the radial Schrödinger equation contains the effective potential

$$V_{\text{eff}}(r) = V(r) + l(l+1)/r^2 \tag{1.14}$$

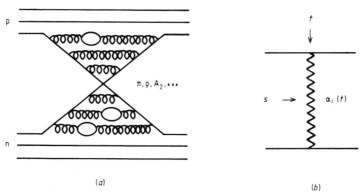

Figure 8. Proton–neutron scattering at small angles, illustrating (*a*) the complicated parton exchange structure and (*b*) the equivalent, but more useful, Regge trajectory exchange, $\alpha_i(t)$ with $i = \pi, \rho, A_2 \ldots$.

where the second term on the right-hand side is the repulsive centrifugal barrier term, which increases with the orbital angular momentum l, and represents the fact that it is harder to bind high l states because of the centrifugal repulsion. This is why high l states are generally heavier than low l ones. In fact, one can solve Schrödinger's equation (or for a more relativistic problem the Bethe–Salpeter equation) for arbitrary values of l, and the mass eigenvalue, m, varies continuously along a trajectory in the l plane. m increases with l, connecting the various physically meaningful solutions which exist for $l = n\hbar$ (n integer $\geqslant 0$). So we anticipate that hadrons will lie on so-called 'Regge trajectories' $l = \alpha(m^2)$ such that if m_i is the mass of meson i, and S_i is its spin, then $S_i = \alpha(m_i^2)$ with $S_i = 0, \hbar, 2\hbar, \ldots$. An example of such a trajectory connecting some of the lighter meson states is shown in figure 9.

As we shall find in chapter 7, Regge theory also predicts that the high-energy behaviour of a hadron scattering amplitude at small angles will take the form $A(s, t) \sim s^{\alpha(t)}$ (where s is the square of the centre-of-mass energy, and $-t$ is the square of the momentum transferred) if $\alpha(t)$ is the trajectory of the particles which can be exchanged (figure 8(*b*)). This is the modern generalisation of Yukawa's meson exchange

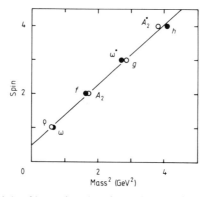

Figure 9. Chew–Frautschi plot of (approximately exchange-degenerate) meson Regge trajectories. $\bigcirc$, $I = 1$; $\bullet$, $I = 0$.

hypothesis, and is very successful in explaining hadron scattering cross sections. Regge theory thus incorporates many of the most complicated confinement aspects of QCD and hence is an essential tool of particle physics. Its relation to parton ideas will be explained in chapters 7–9.

1.6 Outline of the book

Our aim is to give a fairly elementary account of hadron reaction mechanisms from a modern viewpoint, incorporating the ideas invoked in this introduction.

In the next chapter we review briefly the main features of hadron scattering data which we want to try to understand. Then in chapter 3 we examine in more detail processes in which hadrons are produced with a large transverse momentum (p_T) in which the effects of the point-like quark and gluon interactions are seen rather directly. Chapter 4 is concerned with the corrections with are necessary to make the simple parton picture more compatible with QCD and confinement, while in chapter 5 we look at the jets of hadrons which occur in such processes. In chapter 6 we explore the application of parton ideas to low p_T hadron scattering, while in the following chapter we show how the successes of Regge theory may be incorporated into the picture. This theory is used to explore inclusive hadron production and diffractive scattering in chapters 8 and 9, respectively, and we attempt to bring out the relation between the older Regge theory and the newer parton model approach. Then in chapters 10–13 we show the application of these ideas to collider data on pp and p̄p scattering, discussing the total and elastic cross sections, particle production, jets etc, and the production of the weak bosons $W^{\pm}$, Z^0 and of heavy quark states. Chapter 12 includes a brief résumé of the electroweak gauge theory needed to describe W and Z physics at the CERN p̄p Collider. We conclude with a summary in chapter 14.

Because we are trying to cover a very large field of physics our discussion is necessarily somewhat superficial, but this seems appropriate for an introductory survey aimed at readers who are not particularly familiar with the recent developments in particle physics. We have made no attempt to give credit for particular discoveries, but have concentrated on providing useful references, mainly to review articles, which not only contain much more detailed accounts of the individual topics, but also references to the original literature, see, for example, Feynman (1972), Close (1979), Reya (1981), Aitchison and Hey (1982), Altarelli (1982), Leader and Predazzi (1982), Wilczek (1982), Pennington (1983), Halzen and Martin (1984).

2

HIGH-ENERGY HADRON SCATTERING

Most of our information about the structure and properties of hadrons and their constituent partons is based on the analysis of high-energy scattering data. Lepton–hadron scattering is used to probe the weak and electromagnetic structure of hadrons, while hadron–hadron collisions tell us about the more complicated strong interaction properties. In this chapter we want to review, very briefly, the basic features of hadron scattering experiments. Readers requiring more detailed information might look at, for example, Barger (1974), Irving and Worden (1977), Rushbrooke and Webber (1978), Giacomelli and Jacob (1980), Darriulat (1980), Ganguli and Roy (1980) and McCubbin (1981).

2.1 Hadron scattering data

Hadron scattering experiments are performed in two main types of facilities. There are fixed-target experiments in which high-energy beams of hadrons—usually consisting of p, $\bar{p}$, $\pi^{\pm}$ or $K^{\pm}$ but also, less commonly, n, K^0, Σ or Λ—strike particles in a fixed material target (usually protons). If we call the beam particle A, while B is the target particle, their four-momenta will be (setting $c \equiv 1$)

$$P_{A_L} = (E_L, \boldsymbol{p}_L) \qquad p_{B_L} = (m_B, \boldsymbol{0})$$

where E_L and $\boldsymbol{p}_L$ are the laboratory-frame energy and momentum of A and where B, which is at rest, has mass m_B. As a Lorentz-invariant measure of the energy available in the scattering process we introduce

$$s \equiv (p_A + p_B)^2 = (E_L + m_B, \boldsymbol{p}_L)^2 = m_A^2 + m_B^2 + 2m_B E_L \tag{2.1}$$

where in the last step we have used the mass-shell relation

$$E^2 = \boldsymbol{p}^2 + m^2. \tag{2.2}$$

Also there are colliding-beam facilities in which two stored beams of stable hadrons (usually p or $\bar{p}$, but sometimes $\bar{d}$, etc) can be made to collide head on. In this case, if A and B have equal and opposite momenta, so that their four-momenta are

$$p_A = (E_A, \boldsymbol{p}) \qquad p_B = (E_B, -\boldsymbol{p}) \tag{2.3}$$

then the invariant

$$s = (E_A + E_B, \boldsymbol{p} - \boldsymbol{p})^2 = (E_A + E_B)^2 \tag{2.4}$$

is just the square of the total centre-of-mass energy. A comparison of (2.1) and (2.4) indicates the energy ($\sqrt{s}$) gain of colliding beams as compared to fixed-target machines. Their disadvantage is the much lower collision rates achieved with two comparatively tenuous beams.

The accelerators now available range from the older fixed-target machines with E_L of a few GeV to the CERN intersecting storage rings (ISR) in which p beams with energies up to about 30 GeV collide permitting $s \simeq (2 \times 30\ \text{GeV})^2 = 3600\ \text{GeV}^2$. The newly-constructed CERN $\bar{p}p$ collider is now yielding data at energies of about $\sqrt{s} = 540\ \text{GeV}$, and colliders are being constructed at other laboratories to reach $\sqrt{s} = 2000\ \text{GeV}(\equiv 2\ \text{TeV})$ or more.

At the high energies so far available it is found that the total scattering cross section σ_T (for pp $\rightarrow$ X where X represents all types of particles) varies very little with energy, and that $\sigma_T(\text{pp}) \simeq 40\ \text{mb} = (2\ \text{fm})^2$ just as one would expect on geometrical grounds for particles of size about 1 fm (see figure 10). This constancy is an example of 'scaling', i.e. the magnitude of σ_T is independent of the energy scale. However, we see from figure 10 that most total cross sections fall a bit with increasing energy at low energies, and rise slowly at high energies. We shall explore the reasons for this in chapters 9 and 10.

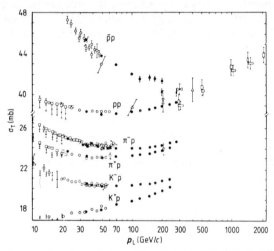

Figure 10. The total cross sections for scattering on protons, taken from Carroll *et al* (1976). More recent data on pp and $\bar{p}p$ are given in figure 86.

At high energies the final state is often quite complicated, several particles being produced in a collision, some of which are likely to miss any detectors which have been set up. It has thus become common to measure the so-called 'inclusive' single-particle production cross section, for say AB $\rightarrow$ CX, where C is the detected particle and X symbolises all the other particles which may have been produced but which have not been observed. The invariant single-particle distribution is defined by

$$f(\text{AB} \rightarrow \text{CX}) \equiv E_C \frac{\text{d}^3\sigma}{\text{d}^3\boldsymbol{p}_C} = \frac{E_C}{\pi} \frac{\text{d}^2\sigma}{\text{d}p_L\,\text{d}p_T^2} \tag{2.5}$$

where $\text{d}^3\sigma/\text{d}^3\boldsymbol{p}_C$ is the differential cross section (i.e. the probability per unit incident flux) for detecting particle C within the phase-space volume element $\text{d}^3\boldsymbol{p}_C$. E_C is included in (2.5) to ensure the Lorentz invariance of f (see, for example, Collins 1977, p327); p_L and p_T are the components of $\boldsymbol{p}_C$ along and transverse to the beam direction, respectively.

A comparatively small fraction of the incident energy goes into making new particles. Thus even at the highest CERN ISR energies the average number of charged particles produced in pp collisions is $\langle n_{ch} \rangle \simeq 12$, 90% of which are pions. The variation with s (in GeV^2) is very approximately (see figure 11(a))

$$\langle n_{ch} \rangle = 2 \log s - 4 \qquad (2.6)$$

whereas if a fixed fraction of the energy went into particle production we would find $\langle n_{ch} \rangle \sim s^{1/2}$.

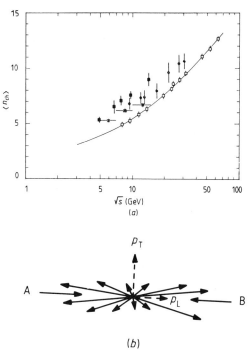

(a)

(b)

Figure 11. (a) The energy dependence of the averaged charged multiplicity for pp collisions, compared with that for e^+e^- annihilations, $\bar{p}p$ annihilations and νp interactions, taken from Berger *et al* (1980). ●, e^+e^-(PLUTO), $K_s^0 \to \pi^+\pi^-$ excluded; □, pp interactions; ■, $\bar{p}p$ annihilation; *, νp interactions. (b) Schematic drawing of AB scattering producing a two-jet event and showing the limited transverse momentum of the produced particles.

Instead, most of the incident energy emerges as the kinetic energy of the outgoing particles, many of which are found to be travelling in two jets with only a rather small deviation from the directions of the incoming particles. Figure 11(b) shows a typical plot of the momenta of the outgoing particles and it will be evident that whereas the longitudinal momentum component, p_L, takes a wide range of values, the transverse momentum, p_T, is generally quite small (<1 GeV/c). The production cross section falls with increasing p_T roughly like (figure 12(a))

$$\frac{d\sigma}{dp_T} \sim \exp(-6p_T) \qquad (2.7)$$

giving an average p_T of only about 0.35 GeV/c, though this does increase slowly with energy.

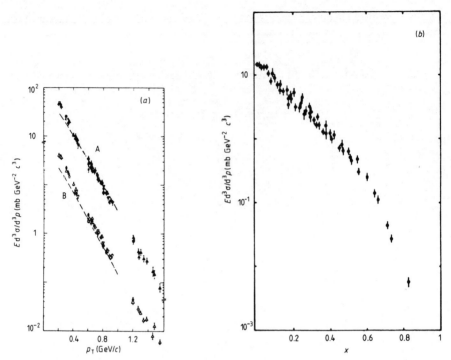

Figure 12. The invariant inclusive cross section as a function of (a) p_T at fixed x ($y_{lab} = 1.5$) (A, $p+p \to \pi^+ + \cdots$; B, $p+p \to \pi^- + \cdots$), and (b) x at fixed p_T (0.4 GeV/c) (see Giacomelli and Jacob 1979).

It is convenient to introduce the dimensionless variable

$$x \equiv p_L/p \qquad (2.8)$$

which measures, in the centre-of-mass frame, the fraction of the beam's momentum (p) which is contained in the longitudinal momentum component (p_L) of the detected particle. Clearly x varies from -1 to 1. A typical example of a cross section plotted as a function of x is shown in figure 12(b), from which it will be seen that a fairly large number of slow particles is produced (x near 0), but that the distribution decreases rapidly to zero as $x \to 1$, like $(1-x)^n$. The shape of this x distribution is found to be essentially independent of energy, an effect which is generally known as 'Feynman scaling' (Feynman 1969). The value of the exponent, n, in these distributions will be of great interest to us in chapter 6.

Another variable which is often used to display the p_L dependence of the cross section is the rapidity, y, defined by (De Tar 1971)

$$y = \tfrac{1}{2} \log \left(\frac{E + p_L}{E - p_L} \right). \qquad (2.9)$$

Clearly y depends on the choice of frame, but it has the advantage of being simply additive under Lorentz boosts along the z axis. Thus if we consider a frame boosted by velocity u, so that

$$E \to \gamma(E + up_L) \qquad p_L \to \gamma(p_L + uE)$$

where $\gamma = (1 - u^2)^{-1/2}$, then we see that

$$y \to y + \tfrac{1}{2} \log \left(\frac{1+u}{1-u} \right) \equiv y + y_{\text{boost}}.$$

In the non-relativistic limit ($v \ll 1$) $E \to m$, $p \to mv$, and thus $y \to v$.

The cross section for inclusive particle production as a function of the centre-of-mass rapidity is shown in figure 13, which exhibits a central plateau at small y and falling cross sections in the fragmentation regions where $y \to \pm y_{\text{max}}$. The magnitude of the central region cross section changes only rather slowly with energy, indicating an approximate scaling, which we will discuss further in chapter 8.

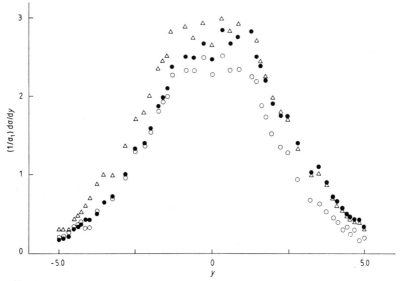

Figure 13. The inclusive rapidity distribution in pp collisions. $\bigcirc$ p_{ISR} 15.4 GeV/c, $\triangle$ p_{ISR} 26.7 GeV/c, $\bullet$ beam 1 15.4 GeV/c, beam 2 26.7 GeV/c.

Sometimes experiments are designed to detect all the particles in the final state (such as $AB \to CD$, $AB \to CDE$, etc). These are called 'exclusive' experiments because care is taken to ensure, using energy, momentum and quantum number conservation arguments, that the detected particles were indeed the only ones produced in the event. As the energy increases any particular exclusive final state contributes only a diminishing fraction of the total cross section.

For future convenience we introduce here the kinematic invariants needed to describe two-body final-state processes of the form $AB \to CD$, shown in figure 14.

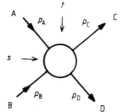

Figure 14. Kinematic variables for the process $AB \to CD$.

We have already introduced ((2.1))

$$s \equiv (p_A + p_B)^2$$

the total centre-of-mass energy squared. We shall also need the momentum transfer variable:

$$t \equiv (p_A - p_C)^2$$
$$= m_A^2 + m_B^2 - 2p_A \cdot p_C \qquad (2.10)$$
$$= m_A^2 + m_B^2 - 2E_A E_C + 2|p_A||p_C| \cos\theta$$

where $p_A \equiv (E_A, p_A)$ is the four-momentum of particle A (and similarly for C) and θ is the scattering angle in the centre of mass frame. It is also useful to define

$$u \equiv (p_A - p_D)^2 \qquad (2.11)$$

but four-momentum conservation requires that

$$s + t + u = m_A^2 + m_B^2 + m_C^2 + m_D^2 \qquad (2.12)$$

so only two of s, t and u are independent variables. A more complete discussion of all these variables can be found, for example, in Collins (1977) or Ganguli and Roy (1980).

2.2 Scattering processes from a parton viewpoint

Nowadays we regard hadrons as clusters ('bags') of confined partons (i.e. quarks, antiquarks and gluons). The scattering of two hadrons, A and B, may thus be viewed as in figure 15. Though each cluster is colourless overall, within it there will be a distribution of colour charge, and the approach of the other cluster will induce a redistribution (polarisation) of this colour charge (just like the polarisation of an atom's electron distribution which is induced by a passing charged particle).

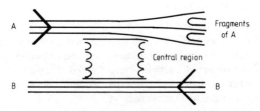

Figure 15. The parton description of a two-jet event, showing the fragmentation of A, central production, and a fast moving particle B little affected by the interaction. In a typical two-jet event B would also fragment, as in figure 11(b).

Some of the partons will presumably be rather little affected, so in the figure many of A's partons travel more or less straight on, but because of the excitation they form into new colourless hadrons, which we call the fragments of A. Some of these hadrons may be carrying a significant fraction of the momentum of A and so appear at large positive values of x (near 1). The incoming partons may be so little influenced that they recombine into the same hadron, as with B in figure 15. In this case we will find in the final state a 'leading particle' with x near -1, which is in fact just the incident hadron B with slightly reduced momentum. Or the incoming hadron may be excited to a more massive state with the same quantum numbers which subsequently decays, such as $p \to N^* \to p\pi$. This is called 'diffraction scattering', and the π is a fragment of the incoming p.

Those partons in A and B which happen to slow down, or to be travelling very slowly in the centre of mass at the moment of collision, may combine to produce new hadrons, nearly at rest. It is this central particle production (at $x \simeq 0$, or $y_{cm} \simeq 0$) which populates the central region plateau. Because most of these particles are pions (the pion is the lightest hadron and hence it occurs in the decay products of most hadrons) this is sometimes called the 'pionisation' region (in analogy with the ionisation produced in the scattering of charged particles).

These high-energy hadron scattering events thus mostly result in two jets of fast moving particles; one containing the fragments of A and the other the fragments of B, together with a central region of low momentum particles not particularly associated with A or B.

This gives us an intuitive understanding of the scaling exhibited by hadron cross sections. The total cross section depends mainly on the geometrical size of the clusters since the effective range of the colour polarisation force does not change much with energy. Similarly, the inclusive cross section for producing a particle in one of the fragmentation regions depends on how the excited parton clusters turn into jets of hadrons, which is essentially independent of the energy of the scattering process. In the central region the cross section increases roughly proportionally to the available rapidity ($y \sim \log s$, hence $\langle n \rangle \sim \log s$) but the cross section per unit rapidity interval changes only slowly with about two charged particles per unit of y (see (2.6)).

Elastic scattering is the special case in which both the incoming clusters of partons retain their integrity and are not broken up to form new hadrons. This is clearly only likely if the collision is rather soft, i.e. there is very little momentum transferred by the energy-independent colour polarisation force. We can therefore expect that the elastic differential cross section, $d\sigma/dt$, will change only slowly with energy and will fall off rapidly with increasing $|t|$. Both of these expectations are vindicated by the data, as we shall find in chapter 9. Very similar remarks apply to quasi-elastic diffractive processes such as $\pi N \to \pi N^*$. The differential cross section does not change much with energy and has a similar t dependence to elastic scattering.

More interesting are two-body final-state processes in which there is an exchange of flavour quantum numbers, such as $\pi^- p \to \pi^0 n$ or $\pi^- p \to K^0 \Lambda$. In the former there is an exchange of charge between the incoming particles, while the latter requires the exchange of both charge and strangeness. Unlike the processes we have been discussing so far, this clearly cannot be achieved just by colour polarisation but, since the flavour quantum numbers are carried by quarks, it is necessary for quarks to pass from one cluster to another, thus reversing their directions of motion. This is hard to achieve at high energies and it is not surprising therefore that the cross section for such processes should fall with increasing energy $\sim s^{-n}$ with typically $n = 1\text{--}2$.

As we mentioned in chapter 1 the two exchanged valence quarks can carry with them a sea of virtual gluons and $q\bar{q}$ pairs, which together comprise a Regge trajectory of hadrons, and we shall find in chapter 7 that the differential cross section has the asymptotic behaviour with energy:

$$d\sigma/dt \sim s^{2\alpha(t)-2} \tag{2.13}$$

where $\alpha(t)$ is the exchanged trajectory, as in figure 8. Again, the occurrence of just two hadrons in the final state is unlikely if the momentum transfer is large and so $d\sigma/dt$ falls rapidly with $|t|$ for this type of process too.

It will be evident from figure 16 that the exponential fall, (2.7), of the inclusive cross section with p_T changes at $p_T \simeq 1\ \mathrm{GeV}/c$ and that although the cross section for large p_T is very small (only about 1 particle in 10^5 has $p_T > 2\ \mathrm{GeV}/c$) it is much larger than one would expect if one simply extrapolated the small p_T behaviour. This change suggests that a different process is involved in large p_T particle production.

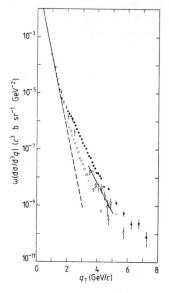

Figure 16. The pioneer ISR data, shown at the 1972 Chicago Conference, on the inclusive production of π^0 mesons at 90° in pp collisions. The broken line is the extrapolation of $\exp(-6p_T)$ which describes the $p_T \lesssim 1\ \mathrm{GeV}/c$ data. $\times\ s = 2850\ \mathrm{GeV}^2$, $\bullet\ s = 2850\ \mathrm{GeV}^2$, $\circ\ s = 950\ \mathrm{GeV}^2$, $\square\ s = 2000\ \mathrm{GeV}^2$.

The transverse momentum, p_T, is the conjugate variable to the impact parameter of the collision b (figure 17(a)) so the large p_T implies that the basic scattering process has occurred at small b. In the parton picture this means that two of the partons have passed very close to each other and so have been scattered at wide angles. These partons attempt to leave the confinement region and in so doing produce jets of hadrons as described in chapter 1. The remaining partons continue almost undisturbed, and so large p_T hadrons will occur in four-jet events, like figure 17. There are the forward and backward jets from the fragmentation of the unscattered partons (similar to those in small p_T scattering) and a pair of almost back-to-back wide-angle jets stemming from the hadronisation of the scattered partons. (The way jets originate from wide-angle parton scattering is discussed in detail in chapter 5.)

Though comparatively rare this type of event is very important because it can give a rather direct insight into the fundamental parton interaction, unlike the more commonplace small p_T events which involve whole clusters of partons. It is for this reason that we shall begin our detailed examination of hadron scattering mechanisms in the next chapter with a discussion of large p_T processes.

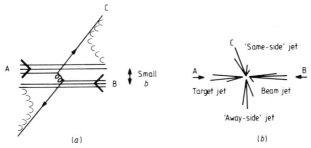

Figure 17. (a) A parton–parton interaction leading to a large p_T AB→CX event, and (b) an idealised picture showing the wide-angle jets arising from the fragmentation of the partons.

First, however, we wish to look more closely at the scaling predictions of the parton model.

2.3 Scaling

As an example of a process which exhibits scaling we consider first the QED reaction $e^+e^- \to \mu^+\mu^-$ (figure 18(a)). In the lowest-order single-photon-exchange approximation the cross section for unpolarised electrons at high energies ($s \equiv (p_{e^+} + p_{e^-})^2 \gg m_e^2$) is (Jauch and Rohrlich 1955, Bjorken and Drell 1964, Itzykson and Zuber 1980)

$$\sigma(e^+e^- \to \mu^+\mu^-) = \frac{1}{s}\left(\frac{4\pi\alpha^2}{3}\right)\left(1 - \frac{4m_\mu^2}{s}\right)^{1/2}\left(\frac{2m_\mu^2 + s}{s}\right) \tag{2.14}$$

$$\xrightarrow[s \gg m_\mu^2]{} \frac{4\pi\alpha^2}{3s}. \tag{2.15}$$

In this expression the first factor s represents the flux of electrons (remember $\sigma \equiv$ the scattering probability per unit flux), the first bracket gives the basic strength of the interaction in first order ($\propto \alpha^2$), the second bracket contains the threshold behaviour and vanishes at the threshold energy $s = 4m_\mu^2$, while the final bracket represents the helicity structure and contains two terms because the outgoing μ may have parallel or antiparallel spins. At high energies these last two factors become irrelevant, however, and we see from (2.15) that

$$s\sigma \to 4\pi\alpha^2/3 \tag{2.16}$$

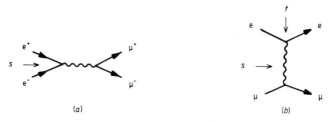

Figure 18. Lowest-order diagrams for (a) $e^+e^- \to \mu^+\mu^-$ and (b) $e\mu \to e\mu$.

independent of s. That is to say, $s\sigma$ is 'scale-invariant', or 'scales'. Since the cross section has the dimensions of an area, and we are working in units where $\hbar c = 0.197 \text{ GeV fm} \equiv 1$, an area in fm^2 must be expressed in GeV^{-2}. In (2.15) this dimensional factor comes from the flux factor, s^{-1}, while the interaction appears just through the dimensionless α (though there is scale-breaking s dependence at lower energies in (2.14)).

Similarly for $e\mu \to e\mu$ scattering of figure $18(b)$ the differential cross section is, for $s, t \gg m_\mu^2$ (Close 1979, p 186),

$$\frac{d\sigma}{dt} = \frac{4\pi\alpha^2}{s^2}\frac{1}{2}\left(\frac{s^2+u^2}{t^2}\right) \tag{2.17}$$

where $s + t + u = 2m_e^2 + 2m_\mu^2$ from (2.12) and where the t^{-2} arises from the photon propagator. We thus find that

$$s^2\frac{d\sigma}{dt} = 2\pi\alpha^2\left(\frac{s^2+u^2}{t^2}\right) \xrightarrow[s,t\to\infty]{} f\left(\frac{t}{s}\right) \tag{2.18}$$

is a function only of the dimensionless ratio t/s, and is scale-invariant. Again the flux factor s^{-2} carries the dimensions of $d\sigma/dt$ in (2.17).

There is no scale factor in these expressions because electrons and muons are point-like fundamental particles which have no charge radius to introduce a size scale into the problem. If it were possible to scatter electrons off free quarks we would expect the cross section to be identical to (2.17) (but with $\alpha \to \alpha e_q$ where e_q is the quark charge in units of e) because a quark is also a point-like fundamental particle.

However, if we scatter an electron off a meson with a large transfer of momentum, t, the simplest parton diagram we can draw is figure $19(a)$ in which the electron scatters off one of the quarks. We then have a $q\bar{q}$ system in which the constituents are moving apart rapidly, and the chances of them pulling back together via gluon exchange to re-form the meson is obviously quite small. In fact, the probability amplitude should be proportional to the meson form factor, $F_M(t)$, which goes like t^{-1} at large t, since $t = q^2$ represents the amount by which the struck quark is off its mass shell after the electron has hit it. To see this recall that in Feynman perturbation theory a particle of mass m and four-momentum k has a propagator $\sim(k^2-m^2)^{-1}$,

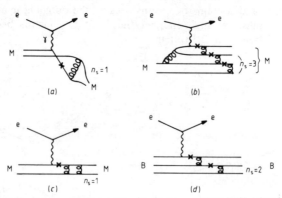

Figure 19. Parton diagrams for large momentum transfer electron–meson (a), (b), (c) and electron–baryon (d) elastic scattering. Internal quark lines far off mass shell are indicated with a cross. In (a) there is one spectator quark to the electron quark scattering, while in (b) there are three, and two in (d).

and $k^2 - m^2$ is the amount by which the particle is off its mass shell (2.2). In figure 19(a) the quark line is off its mass shell as it travels between the photon vertex and the gluon vertex. Of course the internal photon and gluon lines also represent off-shell particles but as we have seen in (2.18) the spin-averaged amplitude for massless vector exchange between massless point-like fermions is scale invariant because the t^{-1} in the propagator is cancelled by the t-dependence of the coupling in the numerator. Hence only the internal off-shell quark lines matter in determining the t-dependence of the full amplitude. See Sivers *et al* (1976, p 64) for a more detailed discussion.

Hence for eM → eM we have at large s and $t(\gg m_M^2)$

$$s^2 \frac{d\sigma}{dt} = 4\pi\alpha^2 \frac{1}{2} \frac{us}{t^2} F_M^2(t) \xrightarrow[s,t \to \infty]{} f\left(\frac{t}{s}\right) F_M^2(t) \sim f\left(\frac{t}{s}\right) t^{-2}. \qquad (2.19)$$

The different angular factors in (2.18) and (2.19) $(1 + u^2/s^2$ and $u/s)$ simply arise because the electron is scattering on a spin-$\frac{1}{2}$ muon as compared to a spin-0 meson. The form factor, $F_M(t)$, is a measure of the non-point-like structure of the meson. It is found that, for example, the pion form factor is well represented by

$$F_\pi(t) = \left(1 - \frac{t}{m_\rho^2}\right)^{-1} \underset{t \to \infty}{\sim} t^{-1} \qquad (2.20)$$

where m_ρ, the mass of the ρ meson, introduces a mass scale into the problem. Form (2.20) is expected according to the vector dominance hypothesis in which the photon couples through the lightest available vector mesons (see Gilman 1972, Gourdin 1974).

We must, of course, also consider many other types of diagram, such as figure 19(b) for example, in which the electron scatters off a sea quark rather than a valence quark. But in this case the number of off-shell propagators, and hence the number of gluons needed to pull the meson back together again, is greater and the amplitude behaves like $\sim t^{-N}$ where N is the number of such propagators. The simpler figure 19(a) will thus be the dominant diagram at large t. There are also multi-gluon exchange diagrams like figure 19(c), but these produce only log (t) modifications to figure 19(a), at least at large t where QCD perturbation theory should be reliable (see Brodsky and Lepage 1979a,b).

For electron–baryon scattering the simplest diagram is figure 19(d) in which again the electron strikes one of the quarks, but now a minimum of two gluons is needed to pull the three quarks back together again. There are thus two off-shell quark propagators and so the baryon's form factor should behave like $F_B(t) \sim t^{-2}$. In fact, the spin-$\frac{1}{2}$ proton has two independent form factors, $G_E(t)$ and $G_M(t)$, which represent its electric and magnetic couplings respectively (see Gourdin 1974, Close 1979) but both are found to behave like

$$G_{E,M}(t) \approx \left(1 - \frac{t}{0.71}\right)^{-2} \sim t^{-2}. \qquad (2.21)$$

Again, more complicated diagrams either give non-leading contributions as $t \to \infty$ or produce only logarithmic modifications to this leading power behaviour.

These arguments are readily generalised, and the rule is that any form factor should have the leading power behaviour as $t \to \infty$

$$F(t) \sim (t)^{-n_s} \qquad (2.22)$$

where n_s is the minimum possible number of 'spectator' partons which are not involved in the initial electron–parton interaction. n_s is equal to the minimum number of gluons required to hold the hadron together, and hence gives the number of far off-shell quark propagators. For mesons $n_s = 1$ while for baryons it is 2. Hence for electron–hadron scattering we have the 'dimensional counting rule' (Sivers *et al* 1976) that

$$s^2 \frac{d\sigma}{dt} \underset{s,t\to\infty}{\sim} f\left(\frac{t}{s}\right)(t)^{-2n_s} \tag{2.23}$$

where $f(t/s)$ is a dimensionless function of the dimensionless ratio t/s, while the factor $(t)^{-2n_s}$ gives the degree of breaking of the scaling law (2.18) for point-like particles, and is characteristic of the composite structure of the hadron.

It should be noted that it is only because we are interested in those unusual events where just a single meson or baryon is found in the final state that we obtain the suppression in (2.23). If instead we consider the total cross section for, say, $ep \to eX$ where X is any number of hadrons then, as we shall discover in chapter 3, one again finds scaling. This is because no far-off-mass-shell propagators are required as the excess energy can be radiated away into hadrons a bit at a time as sketched in figure 20.

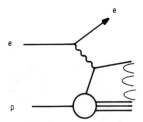

Figure 20. Parton diagram for $ep \to eX$.

Next we look at hadron–hadron scattering. The basic process is qq scattering, and the simplest single-gluon exchange diagram (figure 21), results in (Cutler and Sivers 1978)

$$\frac{d\sigma}{dt}(qq \to qq) = \frac{2}{9} \frac{4\pi\alpha_s^2}{s^2} \frac{1}{2}\left(\frac{s^2+u^2}{t^2}\right) \underset{s,t\to\infty}{\sim} f\left(\frac{t}{s}\right)\frac{1}{s^2} \tag{2.24}$$

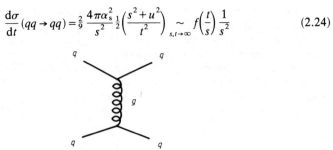

Figure 21. Lowest-order diagram for qq scattering.

just like (2.17) except that $\alpha \to \alpha_s$ and we have acquired a factor $\frac{2}{9}$ from summing over all possible colours of the exchanged gluon (compare (1.7) to (1.9)). At large t where $\alpha_s(t)$ becomes small (see (1.11)) we expect this to be the dominant contribution, but of course in the absence of free quarks it is impossible to check directly this scaling prediction stemming from the point-like nature of the quarks.

If instead we consider meson–meson scattering the simplest diagram is figure 22 in which the basic qq scattering is followed by the same sort of recombination

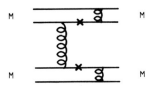

Figure 22. Parton diagram for large momentum transfer MM→MM scattering. The crosses indicate far-off-mass-shell quarks. Of course, the basic hard gluon exchange between the mesons must be supplemented by further soft gluons to ensure that the final mesons are colour singlets.

mechanism as figure 19(a), and so we expect that (Sivers *et al* 1976)

$$\frac{d\sigma}{dt}(MM \to MM) \underset{s,t \to \infty}{\sim} \frac{d\sigma}{dt}(qq \to qq)F_M^4(t) \tag{2.25}$$

$$\sim \frac{1}{s^2}f\left(\frac{t}{s}\right)\left(\frac{1}{t}\right)^4 \tag{2.26}$$

which $\sim s^{-6}$ at fixed values of t/s, i.e. at fixed angle. In this expression we have ignored the contributions of the soft gluons which must also be exchanged to keep the final state M colourless, but which are expected to produce at most logarithmic modifications. Similarly for meson–baryon and baryon–baryon scattering we have

$$\frac{d\sigma}{dt}(MB \to MB) \underset{s,t \to \infty}{\sim} \frac{d\sigma}{dt}(qq \to qq)F_M^2(t)F_B^2(t) \sim \frac{1}{s^2}f\left(\frac{t}{s}\right)\left(\frac{1}{t}\right)^6 \tag{2.27}$$

$$\frac{d\sigma}{dt}(BB \to BB) \underset{s,t \to \infty}{\sim} \frac{d\sigma}{dt}(qq \to qq)F_B^4(t) \sim \frac{1}{s^2}f\left(\frac{t}{s}\right)\left(\frac{1}{t}\right)^8 \tag{2.28}$$

so that at fixed angle these processes should behave like s^{-8} and s^{-10}, respectively. These predictions seem to work rather well for $|t| \gtrsim 2.5$ $(GeV/c)^2$. We are thus led to the more general dimensional counting rule that

$$s^2\frac{d\sigma}{dt}(AB \to CD) \underset{\substack{s,t \to \infty \\ s/t \text{ fixed}}}{\sim} f\left(\frac{t}{s}\right)s^{-2n_s} \tag{2.29}$$

where n_s is the minimum of spectator partons in the basic parton scattering process (which in turn is equal to the number of off-shell quark propagators, or the number of gluon propagators required over and above the scaling gluon exchange of the initial qq scattering). As before lowest-order parton model predictions will acquire logarithmic corrections from higher-order diagrams with further gluon exchanges like those of figure 19(c) (Brodsky and Lepage 1979a,b).

The breakdown of strict scaling in (2.29) stems from the fact that we are demanding the re-formation of a specific two-body final state at wide angles. If instead we allow the struck quarks' energy to be freely radiated away through hadron production we expect that the inclusive production process AB → CX, where C is the detected large

p_T hadron and X represents everything else produced, will take the asymptotic form

$$f(AB \to CX) \underset{s,p_T \to \infty}{\sim} f\left(\frac{p_T}{\sqrt{s}}, \theta_{cm}\right) p_T^{-4} \tag{2.30}$$

where f is some dimensionless function of the dimensionless variables $p_T/\sqrt{s}$ and the centre-of-mass scattering angle, and the p_T^{-4} behaviour arises from the natural scale of the basic qq scattering process (2.24). In other words we anticipate that $p_T^4 f(AB \to CX)$ will be scale-invariant. We shall review the success of this prediction in some detail in chapters 3 and 4 (see Jacob and Landshoff 1978).

Finally we must look again at the total cross section $AB \to X$. We have said that this will be dominated by small p_T processes in which the incoming partons deviate rather little from their initial directions of motion. If we suppose that first a gluon is exchanged between the hadrons, and then particle production occurs to try and neutralise the colour, the optical theorem can be used to perform the sum over all the possible final states as in figure 23. Since the gluon exchange amplitude at fixed t behaves like s as $s \to \infty$ (cf (2.24)) we obtain

$$\sigma_T(AB \to X) \underset{s \to \infty}{\sim} (1/s)s = \text{constant} \tag{2.31}$$

i.e. the total cross section should be invariant as $s \to \infty$. We can anticipate $\log s$ modifications to this result due to multiple gluon exchange, but this rather constant behaviour of total cross sections is certainly in accord with figure 10.

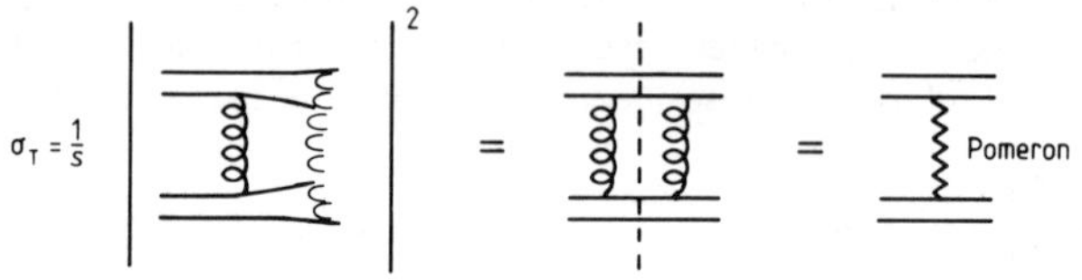

Figure 23. Diagrammatic representation of the optical theorem. The first equality indicates that the AB total cross section is the sum over all possible multiparticle production cross sections. The second equality expresses this as an amplitude product where the broken line implies the sum over all possible intermediate states. Via unitarity, this product is equal to the imaginary part of the forward $AB \to AB$ amplitude (which at high energy is dominated by Pomeron exchange).

Long ago Pomeranchuk (1958) predicted (from entirely different considerations) that total cross sections would approach a constant asymptotic limit, and the Regge trajectory whose exchange ensures this behaviour became known as the Pomeron, with $\alpha_P(t=0) = 1$. In the parton model the Pomeron is identified as a colourless, flavourless multiple (two and more) gluon exchange. We shall pursue this identification further in chapter 9.

3

LARGE p_T PROCESSES

As we remarked in the previous chapter large momentum transfer processes provide an insight into the fundamental parton interactions. Here we introduce the naive parton model, which provides a useful first approximation for such processes.

3.1 The parton model for large p_T scattering

If hadrons are made up of more fundamental parton constituents (i.e. quarks and gluons) it must be possible to describe any hadronic reaction in terms of the interactions of these constituents. But this viewpoint will obviously be most useful for those reactions in which the basic parton scattering process is fairly well separated in time from the more complex confinement effect, which prevents the partons from escaping as free particles and causes them to re-assemble in hadrons. The production of hadrons which have a large momentum component (p_T) transverse to the beam direction is a good example of such a reaction (Feynman 1972, Sivers *et al* 1976, Jacob and Landshoff 1978, Close 1979).

The basic diagram for $AB \rightarrow CX$, where C is the hadron with large $|p_T|$ (say $p_T > 2\,\text{GeV}/c$) while X represents all the other particles in the final state, is shown in figure 24. The incoming particles A and B contain, *inter alia*, partons a and b (respectively) which scatter, producing partons labelled c and d (which may often be the same as a and b) which have a large transverse momentum component q_T. Subsequently hadron C is produced from c via the confinement mechanism. Since q_T is the conjugate variable to the impact parameter of the parton scattering process, large q_T implies that the partons have scattered at a small distance where, according to the arguments of § 1.3, α_s is small. Hence we may reasonably hope that perturbation theory will be applicable.

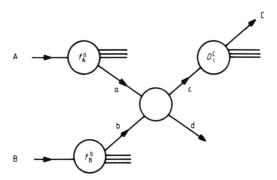

Figure 24. The hadronic interaction $AB \rightarrow CX$ at large p_T in terms of the parton sub-process $ab \rightarrow cd$, the structure functions f_A^a, f_B^b, and the fragmentation function D_c^C.

The uncertainty principle tells us that the time interval τ within which parton scattering occurs is rather short ($\tau \sim \hbar/q_T c$). Partons c and d then fly apart until the confinement mechanism causes hadron production, including the production of C. If we suppose that R is the characteristic distance within which confinement occurs (probably $R \sim 0.5$ fm, a typical hadron 'size') then the time taken for C to be produced will be of the order of $T \sim R q_T/mc^2$ (q_T/mc being the Lorentz dilation in transforming from the C rest frame to the AB centre-of-mass frame). Thus as q_T increases the processes of parton scattering and hadron production become separated by longer times ($T \gg \tau$) and the description given in figure 24 becomes more plausible.

We can in fact use figure 24 to try and estimate the cross section for the inclusive process AB $\rightarrow$ CX. Let $f_A^a(x_a)$ be the probability that hadron A contains a parton a which is carrying a fraction $x_a = q_a/p_A$ of its momentum, $0 \le x_a \le 1$. For the time being we neglect any momentum a may have in directions transverse to the beam direction (the z axis), and also neglect the masses of both hadrons and partons as being small compared to the momenta we are considering, so the four-momenta may be approximated as

$$p_A \simeq (p_A, \mathbf{0}, p_A)$$
$$q_a \simeq (q_a, \mathbf{0}, q_a) = x_a(p_A, \mathbf{0}, p_A) \tag{3.1}$$

since $p_A \gg m_A$, etc. The functions $f_A^a(x_a)$ are called the 'structure functions' of A. Similarly we introduce the 'fragmentation function' $D_c^C(z_c)$ representing the probability that the outgoing parton c produces a hadron C carrying a momentum fraction $z_c = p_C/q_c$, where $0 \le z_c \le 1$. We are assuming that C is produced collinearly with c and that the fragmentation depends only on z_c and is independent of the nature of the initial state.

If we neglect the particle masses, the invariant variables for AB $\rightarrow$ CX are

$$s = (p_A + p_B)^2 \simeq 2p_A \cdot p_B$$
$$t = (p_A - p_C)^2 \simeq -2p_A \cdot p_C \tag{3.2}$$

where $\sqrt{s}$ is the total centre-of-mass energy and $\sqrt{-t}$ is the invariant momentum transfer from A to C. The corresponding variables for the parton sub-process, ab $\rightarrow$ cd, are

$$\bar{s} \equiv (q_a + q_b)^2 \simeq 2q_a \cdot q_b = 2x_a x_b p_A \cdot p_B \simeq x_a x_b s$$
$$\bar{t} \equiv (q_a - q_c)^2 \simeq -2q_a \cdot q_c = -2x_a p_A \cdot p_C/z_c \simeq x_a t/z_c. \tag{3.3}$$

We can express the invariant cross section for AB $\rightarrow$ CX as the weighted sum of differential cross sections, $d\sigma/d\bar{t}$, of all possible parton scatterings that can contribute:

$$E_C \frac{d\sigma}{d^3 p_C}(AB \rightarrow CX) = \sum_{abcd} \int_0^1 dx_a \int_0^1 dx_b\, f_A^a(x_a) f_B^b(x_b) \frac{1}{\pi z_c} \frac{d\sigma}{d\bar{t}}(ab \rightarrow cd) D_c^C(z_c) \tag{3.4}$$

as shown in detail, for example, in the review by Sivers et al (1976); see also Feynman and Field (1977). So we can predict the cross section if we know the structure functions, the fragmentation functions and the cross sections for all the parton sub-processes.

3.2 The structure functions

The best probes for determining the structure functions of nucleons are deep inelastic lepton–nucleon scattering processes such as $eA \to eX$, $\mu A \to \mu X$, $\nu_\mu A \to \mu^- X$ and $\bar{\nu}_\mu A \to \mu^+ X$ with a large momentum transferred between the lepton and the nucleon, A. Thus in electron scattering (figure 25) the cross section for electron–quark scattering is well known (apart from the quark charge it is the same as for $e\mu \to e\mu$) and we can write formally (Gilman 1972)

$$\frac{d^2\sigma}{dx \, dy}(eA \to eX) \propto \sum_q f_A^q(x)\frac{d\sigma}{dt}(eq \to eq). \qquad (3.5)$$

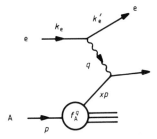

Figure 25. Deep inelastic lepton–nucleon scattering, $eA \to eX$, via photon exchange between the electron and a quark of nucleon A. The particle four-momenta are shown.

Of course the electron 'sees' only the charged partons, i.e. the quarks. In place of the invariant variables $Q^2 \equiv -q^2$ and $\nu \equiv p \cdot q$ we have introduced dimensionless quantities

$$x = Q^2/2p \cdot q \qquad y = p \cdot q/p \cdot k_e \qquad (3.6)$$

where the particle four-momenta are defined in figure 25. x is the fraction of the hadron's momentum carried by the quark (which is assumed to be moving collinearly with A), as can be seen from the mass shell condition for the outgoing quark, $(xp + q)^2 = 0$. We neglect particle masses. Note that

$$1 - y = p \cdot k_e'/p \cdot k_e \simeq -u/s \simeq \tfrac{1}{2}(1 + \cos \theta^*) \qquad (3.7)$$

where θ^* is the centre-of-mass scattering angle. The allowed kinematic region for $eA \to eX$ is therefore $0 \le x, y \le 1$.

In the rest frame of the nucleon, $p = (M, 0, 0, 0)$, we have

$$Q^2 \equiv -q^2 = -(k_e - k_e')^2 = 4EE' \sin^2 \theta/2$$

$$\nu \equiv p \cdot q = M(E - E') \qquad (3.8)$$

$$y = (E - E')/E.$$

Thus for a given incident electron energy E, a measurement of the outgoing electron energy, E', and scattering angle θ determines precisely the x value of the quark from which the electron has scattered, as well as y, which is simply the fraction of the electron's energy transferred to the nucleon.

If quarks have spin $\tfrac{1}{2}$ and a point-like coupling to the photon (just like the electron or muon, except for e_q) then it is straightforward to show that the precise form of

(3.5) is (Gilman 1972, Close 1979, p 172)

$$\frac{d^2\sigma}{dx\,dy}(eA \to eX) = \frac{2\pi\alpha^2}{Q^4}s[(1-y)^2+1]\sum_q e_q^2 x f_A^q(x).\tag{3.9}$$

The most important feature of this prediction is the scaling behaviour, i.e. $f_A^q(x)$ depends on the ratio $x = Q^2/2\nu$ and not on Q^2 and ν individually. This is known as Bjorken scaling. If the quarks were not point-like but had a spatial distribution of charge, the eq interaction would depend on the quark's form factor $F(Q^2)$ and so in (3.9) $f_A^q(x)$ would be replaced by $F_A^q(x, Q^2)$. It was the observation of scaling in deep inelastic scattering which provided the principal motivation for the introduction of the parton model.

A similar analysis can be made for deep inelastic neutrino scattering (Llewellyn-Smith 1972) with the obvious modifications due to the short range and parity violation of the weak interaction, and the occurrence of the weak Fermi coupling constant G instead of e^2. For example, for neutrino scattering on a target which contains an equal number of neutrons and protons, the cross section per nucleon is

$$\frac{d^2\sigma}{dx\,dy}(\nu A \to \mu^- X) = \frac{G^2 s}{2\pi}x\sum_{q,\bar{q}}[f_p^q(x)+(1-y)^2 f_p^{\bar{q}}(x)]\tag{3.10}$$

while for $\bar{\nu}$ scattering the proton structure functions are interchanged $f_p^q \leftrightarrow f_p^{\bar{q}}$. The y distribution follows directly from the helicity structure of the neutrino–quark interaction. There are only left-handed q and ν charged-current weak interactions, so the helicities of the interacting particles are both $-\frac{1}{2}$, giving a total spin of the νq system $S_z = 0$. This results in an isotropic (y-independent) cross section. However, the $\bar{q}$ is right-handed (helicity $+\frac{1}{2}$) so the $\nu\bar{q}$ system has $S_z = -1$. For backward scattering this would have to change to $S_z = 1$, which is impossible as the z component of the spin must be conserved, and so we get a $(1-y)^2$ distribution which vanishes for $\theta^* = \pi$ (see (3.7)). The electromagnetic current has no 'handedness' and so both terms appear with equal weight in (3.9).

Another type of process in which hadron structure functions can be studied is high-mass lepton-pair production such as $AB \to \mu^+\mu^- X$. This is known as the Drell–Yan process (Drell and Yan 1971). In the parton model (figure 26) the cross section is (see Stroynowski 1981)

$$\frac{d\sigma}{dm^2}(AB \to \mu^+\mu^- X) = \left(\frac{4\pi\alpha^2}{3m^2}\right)\frac{1}{3}\sum_q e_q^2 \int_0^1 dx_a \int_0^1 dx_b [f_A^q(x_a)f_B^{\bar{q}}(x_b)$$

$$+ f_A^{\bar{q}}(x_a)f_B^q(x_b)]\delta(m^2 - x_a x_b s)\tag{3.11}$$

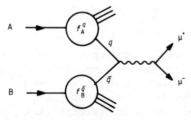

Figure 26. The basic $q\bar{q} \to \gamma \to \mu^+\mu^-$ parton model interaction for the Drell–Yan process $AB \to \mu^+\mu^- X$.

where m is the mass of the lepton pair. The first factor in brackets is the high-energy QED cross section for $e^+e^- \to \mu^+\mu^-$, since $q\bar{q} \to \mu^+\mu^-$ is the same apart from the quark charge. The extra factor $\frac{1}{3}$ accounts for the fact that all three colours of q and $\bar{q}$ occur with equal probability but only a q and $\bar{q}$ of the same colour can annihilate to form a colourless photon.

The same quark structure functions should appear in all the above processes and it is an important test of the parton idea that all these types of data can be accounted for by an identical parametrisation of the structure functions (Fox 1977, Buras and Gaemers 1978, Buras 1980). An example of the resulting fits is given in figure 27.

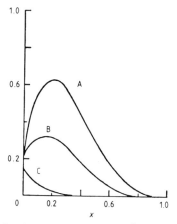

Figure 27. The quark structure functions of the proton, $q(x) \equiv f_p^q(x)$, taken from Barger and Phillips (1974). A, $xu(x)$; B, $xd(x)$; C, $x\bar{u}(x) = x\bar{d}(x) = x\bar{s}(x)$.

Several aspects of these distributions can be deduced from quite elementary considerations. First, the structure functions must be consistent with the quantum numbers of the hadron (charge, isospin, strangeness, . . .). Thus for a proton, whose quantum numbers are those of the uud combination of 'valence' quarks, we have the sum rules

$$\int_0^1 [f_p^u(x) - f_p^{\bar{u}}(x)] \, dx = 2 \qquad \int_0^1 [f_p^d(x) - f_p^{\bar{d}}(x)] \, dx = 1 \qquad \int_0^1 [f_p^s(x) - f_p^{\bar{s}}(x)] \, dx = 0.$$

$$(3.12)$$

The 'sea' quarks appear in $q\bar{q}$ pairs which do not affect the net quantum numbers. Furthermore, between them the partons, each with momentum fraction xp_A, must carry all the momentum of the hadron, p_A, so

$$\sum_a \int_0^1 xf_A^a(x) \, dx = 1. \qquad (3.13)$$

In fact, if one inserts the observed quark distributions in a proton into (3.13) one obtains only about 0.5 on the RHS of (3.13) (CHARM 1981) from which it is concluded that the gluons, which are not seen by the weak or electromagnetic probes, are carrying about half the proton's momentum.

In the limit as $x \to 1$ we would have a single parton carrying all the momentum of the hadron, but this is clearly impossible and so the structure functions must vanish. As this limit is approached all the other spectator quarks must have vanishing momentum and hard gluon exchanges are needed to transfer their momentum to the fast quark, just like the exchanges in figure 19 which lead to (2.23). This results in the dimensional counting rule (Brodsky and Farrar 1973) for the behaviour of the structure function, that

$$f(x) \underset{x \to 1}{\sim} (1-x)^{2n_s - 1} \tag{3.14}$$

where n_s is the minimum number of other (spectator) partons whose momentum would have to vanish in this limit. Thus a valence quark (q_v) in a meson must be accompanied by at least $\bar{q}_v$ so $n_s - 1$ and $f_M^v(x) \sim (1-x)$, while a sea quark (q_s) has at least three spectators and $f_M^s(x) \sim (1-x)^5$. Similarly a gluon has at least two spectators $(q_v \bar{q}_v$, see figure 28) and $f_M^g \sim (1-x)^3$. These counting rules are discussed in more detail in § 6.2.

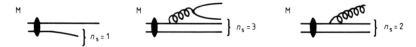

Figure 28. The minimum number of spectators, n_s, accompanying a valence, a sea quark and a gluon in a meson.

From (3.6) we see that as $x \to 0$, $\nu \to \infty$ for fixed q^2, and so in this limit we are in the regime of high-energy virtual-photon–hadron scattering. This is called the 'Regge' regime because in the high-energy limit hadron scattering amplitudes are controlled by the exchange of particles which lie on Regge trajectories, as we mentioned in § 1.5 and will discuss in more detail in chapter 7. The dominant exchanges are the flavourless Pomeron (P) and the leading meson Regge trajectory (R) of figure 9, and it is found that (Gilman 1972) the total γA cross section has the form (cf (7.22))

$$\sigma_T(\gamma A) \underset{\nu \to \infty}{\longrightarrow} \beta_P \nu^{\alpha_P - 1} + \beta_R \nu^{\alpha_R - 1} \tag{3.15}$$

(modulo log ν factors) where β_P and β_R are coefficients, the constant $\alpha_P \simeq 1$ behaviour being independent of the quantum numbers of A, while $\alpha_R(<1)$ is the leading Regge trajectory (i.e. particle exchange) contribution, which does depend on the flavour of A. From (3.9) we deduce that

$$\sigma_T(\gamma A) \propto \sum_q e_q^2 x f_A^q(x) \tag{3.16}$$

assuming that the high-energy region ($\nu \to \infty$ at fixed Q^2) overlaps the scaling region ($\nu, Q^2 \to \infty$ at fixed $x = Q^2/2\nu$). Thus identifying the first term on the RHS of (3.15) with the sea quarks (flavour-independent) contribution, and the second with the valence quarks, we obtain (Drell and Yan 1970, West 1970, Feynman 1972)

$$f_A^s(x) \underset{x \to 0}{\longrightarrow} x^{-\alpha_P} \qquad (\alpha_P \simeq 1)$$

$$\tag{3.17}$$

$$f_A^v(x) \underset{x \to 0}{\longrightarrow} x^{-\alpha_R} \qquad (\alpha_R \simeq \tfrac{1}{2} \text{ for } u \text{ and } d \text{ quarks}).$$

Thus the sea quarks have a bremsstrahlung-like spectrum at small x (gluons creating $q\bar{q}$ pairs) and the number increases logarithmically as $x \to 0$. This behaviour is borne out by comparing $ep \to eX$ and $en \to eX$ data. At small x the parton multiplicity is high and isospin-independent giving the observed $\sigma_p \approx \sigma_n$. At large x there are few partons and the difference of the p, n valence quark composition means $\sigma_p \neq \sigma_n$, as shown by the data.

Combining (3.14) and (3.17) we arrive at the approximate form of the structure function:

$$f_A^a(x) = C_A^a x^{-\alpha} (1-x)^{2n_s - 1} \tag{3.18}$$

though of course this is only the leading behaviour as $x \to 0$ and 1. The normalisation C_A^a can be fixed by the sum rules (3.12) and (3.13). We thus have quite a good idea of what the structure functions should look like, and figure 27 clearly bears out these expectations.

3.3 The fragmentation functions

The final stage of figure 24, the conversion of the high momentum parton c into the hadron C, is supposed to be independent of how c was produced. Hence we can obtain the fragmentation function $D_c^C(z)$ rather directly from the process $e^+e^- \to CX$ (see figure 29) in which the initial state has no hadrons to confuse matters (see Wiik and Wolf 1979).

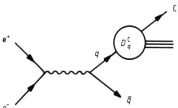

Figure 29. Determination of the fragmentation function, $D_q^C(z)$, from $e^+e^- \to CX$. Hadron C has a fraction z of the quark's momentum.

In the parton model the cross section is given by

$$\frac{d\sigma}{dz}(e^+e^- \to CX) = \sum_q \sigma(e^+e^- \to q\bar{q})[D_q^C(z) + D_{\bar{q}}^C(z)]$$

$$= \left(\frac{4\pi\alpha^2}{3s}\right) 3 \sum_q e_q^2 [D_q^C(z) + D_{\bar{q}}^C(z)] \tag{3.19}$$

where the first bracket is the QED cross section for $e^+e^- \to \mu^+\mu^-$ via the one-photon intermediate state, the additional factor of 3 stems from three different colours of quarks which can be produced, and e_q is the charge of the quark. Clearly C may have been produced from either the q or the $\bar{q}$. Note that

$$z \equiv p_C/q_q \approx p_C/E \tag{3.20}$$

where E is the electron's centre-of-mass energy, since initially the $q\bar{q}$ pair must carry the full e^+e^- energy.

Similarly, in deep inelastic hadron production $lA \to l'CX$ (where l and l' are leptons) the cross section is given by

$$\frac{d^3\sigma}{dx\,dy\,dz} = [(3.9)\text{ or }(3.10)]D_c^C(z) \tag{3.21}$$

provided that C is a fragment of the struck quark c and not of the spectators in figure 25.

The form of the $D_c^C(z)$ is partially determined by sum rules, and by the limiting behaviour as $z \to 0$ and 1.

Thus, since the energy of all the hadrons which fragment from a given quark must equal the initial energy of that quark we have

$$\sum_C \int_0^1 z D_q^C(z)\,dz = 1 \tag{3.22}$$

while charge conservation requires

$$\sum_C e_C \int_0^1 [D_q^C(z) - D_{\bar{q}}^C(z)]\,dz = e_q. \tag{3.23}$$

The average multiplicity of hadrons of type C is given by

$$\langle n_C \rangle = \sum_{q,\bar{q}} \int_{z_{min}}^1 D_q^C(z)\,dz \tag{3.24}$$

where z_{min} is the lowest value of z possible for a hadron of mass m_C. As $z \to 1$ the hadron takes all of the parton's momentum so any other partons which are left behind in the hadronisation must have negligible momentum. Hence dimensional counting leads us to expect that

$$D_q^C(z) \underset{z \to 1}{\sim} (1-z)^{2n_s - 1} \tag{3.25}$$

where n_s is the minimum possible number of spectators. Thus for a quark q fragmenting into a meson we have $n_s = 1$ if M contains q and $n_s = 3$ if it does not; see figure 30.

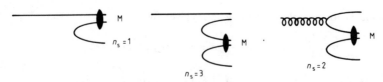

Figure 30. The minimum number of spectators, n_s, accompanying the fragmentation of partons into a meson M.

As $z \to 0$ the essentially massless hadrons are taking none of the parton's momentum and so we expect $D(z) \sim z^{-1}$, which gives a logarithmic increase of $\langle n_C \rangle$ with energy in (3.24). It is thus convenient to approximate

$$D_q^C(z) = D_q^C z^{-1} (1-z)^{2n_s - 1} \tag{3.26}$$

with the normalisation D_q^C bounded by the sum rules (3.22) and (3.23).

3.4 Parton scattering cross sections

The leading-order QCD diagrams for the basic ab → cd sub-process of figure 24 are shown in figure 31. They are very similar to QED diagrams (for example, $qq \to qq$ via gluon exchange is like ee → ee via photon exchange) except for the replacement of α by α_s multiplied by the appropriate SU(3) colour factor. The results are given in table

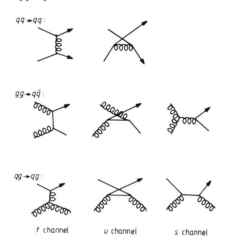

Figure 31. Basic QCD diagrams. Quarks, q, and gluons, g, are shown by straight and curly lines, respectively. The process $q\bar{q} \to q\bar{q}$ has an s channel, but no u channel contribution. In addition there are $gg \to gg$ diagrams.

Table 3. Parton–parton differential cross sections, $d\sigma/dt = \pi\alpha_s^2 |A|^2/s^2$, in lowest order, where here s, t, u are the sub-process variables $\bar{s}$, $\bar{t}$, $\bar{u}$ of (3.3). The initial (final) colours and spins have been averaged (summed). The subscripts 1, 2 denote distinct quark flavours. The table is from Combridge et al (1977).

Subprocess	$|A|^2$
$\left.\begin{array}{l} q_1 q_2 \to q_1 q_2 \\ q_1 \bar{q}_2 \to q_1 \bar{q}_2 \end{array}\right\}$	$\dfrac{4}{9}\left(\dfrac{s^2+u^2}{t^2}\right)$
$q_1 \bar{q}_1 \to q_2 \bar{q}_2$	As above with $s \leftrightarrow t$
$q_1 q_1 \to q_1 q_1$	$\dfrac{4}{9}\left(\dfrac{s^2+u^2}{t^2}+\dfrac{s^2+t^2}{u^2}\right)-\dfrac{8}{27}\dfrac{s^2}{ut}$
$q_1 \bar{q}_1 \to q_1 \bar{q}_1$	As above with $s \leftrightarrow u$
$q\bar{q} \to gg$	$\dfrac{32}{27}\left(\dfrac{u^2+t^2}{ut}\right)-\dfrac{8}{3}\left(\dfrac{u^2+t^2}{s^2}\right)$
$gg \to q\bar{q}$	$\dfrac{1}{6}\left(\dfrac{u^2+t^2}{ut}\right)-\dfrac{3}{8}\left(\dfrac{u^2+t^2}{s^2}\right)$
$qg \to qg$	$-\dfrac{4}{9}\left(\dfrac{u^2+s^2}{us}\right)+\left(\dfrac{u^2+s^2}{t^2}\right)$
$gg \to gg$	$\dfrac{9}{2}\left(3-\dfrac{ut}{s^2}-\dfrac{us}{t^2}-\dfrac{st}{u^2}\right)$

3 (see Combridge *et al* 1977, Cutler and Sivers 1978). The value of α_s is known to be ≈ 0.2 for $Q = 3$ GeV (e.g. from the charmonium spectrum) so higher-order corrections involving higher powers of α_s should not be very important.

With these cross sections, and the results of the previous subsections, we have all the ingredients needed to obtain the cross section for the hadronic process $AB \to CX$ based on (3.4). But even without a detailed knowledge of the structure and fragmentation functions we can predict the basic form of this cross section. Both f and D are dimensionless functions of dimensionless variables (x and z respectively), so the only scale-dependent factors in (3.4) are the parton differential cross sections of table 3, all of which have the fixed-angle form

$$\frac{d\sigma}{d\hat{t}} (ab \to cd) \sim \frac{1}{\hat{s}^2} = GeV^{-4}. \tag{3.27}$$

Hence we anticipate that

$$E_C \frac{d^3\sigma}{d^3p_C} (AB \to CX; s, p_T; \theta_{cm}) = F(\theta_{cm}, x_T) p_T^{-4} \tag{3.28}$$

where $x_T \equiv 2p_T/\sqrt{s}$ is dimensionless (as of course is θ_{cm}) and F is a scale-invariant function. But we have already remarked that the $pp \to \pi^0 X$ cross section behaves like p_T^{-8} for $2 < p_T < 6$ GeV/c, bending to perhaps p_T^{-6} for $p_T > 10$ GeV/c (see figure 16). Clearly the parton model which worked so well for lepton scattering processes has failed us. However, a more general scaling parametrisation

$$E_C \frac{d^3\sigma}{d^3p_C} \sim p_T^{-n} (1 - x_T)^m$$

with n and m arbitrary, can account for much of the data on large p_T scattering (Sivers *et al* 1976). This is indicative of an underlying hard scattering process and suggests that it would be premature to discard the whole idea.

4

SCALING VIOLATIONS

As the basic framework outlined in the previous chapter seems so plausible it is natural to hope that the incorrect prediction, (3.28), is due to the inadequacy of the approximations rather than the fundamental ideas. The next step is thus to try and include higher-order corrections in α_s, the strong coupling constant, to see if these can assist us to better agreement with experiment.

4.1 The coupling constant

We already know one source of violation of the scaling behaviour. We mentioned in chapter 1 that α_s is not a constant but a logarithmically decreasing function of Q^2 due to the (anti-) shielding of the colour charge. The result, (1.11),

$$\alpha_s(Q^2) = \frac{4\pi}{b_0 \log (Q^2/\Lambda^2)}$$

depends on a free parameter, Λ, which determines the scale at which α_s becomes large, and hopefully results in the confinement of quarks within hadrons. Experiments suggest that $0.1 \leqslant \Lambda \leqslant 0.5 \, \text{GeV}^2$ so α_s should certainly be decreasing with $Q^2 = \bar{t}$. It is claimed that this can reduce the effective power of p_T by as much as one unit in the range $2 < p_T < 6 \, \text{GeV}/c$, though this is critically dependent on the value chosen for Λ within the allowed range (Feynman *et al* 1978). The replacement $\alpha_s \rightarrow \alpha_s(Q^2)$ is actually required by the factorisation theorem which we discuss in the next subsection.

4.2 The evolution of the structure and fragmentation functions

In the parton model of chapter 3 we supposed that each hadron had some definite fixed parton structure described by the structure functions $f_A^a(x)$, which depend only on x. But in QCD each parton can radiate other partons, as in figure 32, so the actual number of partons we see depends on the 'resolving power' of the observing system we employ.

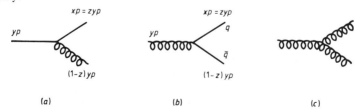

Figure 32. Basic parton processes described by the 'splitting' probability functions (a) $P_{q \rightarrow q}(z)$ (or $P_{q \rightarrow g}(1-z)$), (b) $P_{g \rightarrow q}(z)$ and (c) $P_{g \rightarrow g}(z)$.

39

In electron–proton scattering, for example, the uncertainty principle tells us that the distance resolution is $\Delta x \simeq \hbar/\sqrt{Q^2}$ where $\sqrt{Q^2}$ is the invariant mass of the virtual photon. Thus at low $Q^2 (\ll 1\,\text{GeV}^2)$ an ep scattering experiment will do little more than determine that the proton has a charge and magnetic moment (figure 33(a)). At higher Q^2 ($\lesssim 1\,\text{GeV}^2$) a photon 'sees' the virtual pion cloud (figure 33(b)), and the

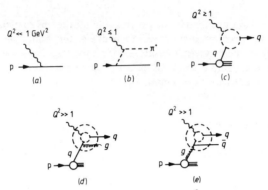

Figure 33. The proton as seen by a virtual photon as Q^2 increases, $(a) \rightarrow (d), (e)$.

proton appears as an extended composite object described by the form factors $G_{\text{E,M}}(Q^2)$. At $Q^2 \gtrsim 1\,\text{GeV}^2$ we begin to see evidence for the three point-like valence quarks (figure 33(c)). If the quarks were non-interacting no further structure would be resolved as Q^2 increased and exact scaling (described by $f_p^q(x)$) would set in. However, due to the basic QCD vertices of figure 32, on increasing the resolution $(Q^2 \gg 1\,\text{GeV}^2)$ we find that each quark is itself surrounded by a gluon cloud and a sea of $q\bar{q}$ pairs. Hence the number of partons which share the hadron's momentum increases with Q^2. There is an increased probability of finding a quark at small x, and a decreased chance of finding one at high x, because the high-momentum quarks lose momentum by radiating gluons. Hence the structure functions evolve with Q^2 as sketched in figure 34, the total area under the curve remaining fixed to ensure that momentum is conserved.

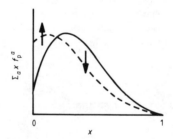

Figure 34. Qualitative pattern of the change of the proton's structure functions with increasing Q^2. Note the increase in the sea quark contribution to xf_p^a at $x = 0$ as Q^2 increases. The area under the two curves is the same. ——, Q_0^2; – – –, Q^2 $(Q^2 > Q_0^2)$.

To determine the Q^2 evolution of structure functions from QCD we follow the physically intuitive approach developed by Altarelli and Parisi (1977) which deals directly with the probability that extra partons are emitted. The alternative more formal approach which starts from the amplitudes for emitting partons, based on renormalisation group techniques, can be found, for example, in the reviews by Buras (1980), Reya (1981) and Bassetto *et al* (1983).

We begin by supposing that the proton consists of valence quarks only, with structure functions $f_p^q(x)$ which give the probability of finding a quark with a fraction x of the proton's momentum, p. Next we consider the possibility that a quark carrying momentum yp emits a gluon, leaving itself with momentum xp ($x \leqslant y \leqslant 1$) as in figure $32(a)$. The probability for this is

$$\frac{\alpha_s(Q^2)}{2\pi} P_{q \to q}\left(\frac{x}{y}\right) \log\left(\frac{Q^2}{Q_0^2}\right) \tag{4.1}$$

to first order in the colour coupling α_s. The splitting function $P_{q \to q}(z)$ is similar to that for $e \to e\gamma$ in the QED theory of bremsstrahlung, and can be calculated from the Feynman rules (Altarelli and Parisi 1977) to be

$$P_{q \to q}(z) = \frac{4}{3} \frac{1 + z^2}{1 - z}. \tag{4.2}$$

The divergence as $z \to 1$ is cancelled when virtual gluon contributions are included. The $\log Q^2$ arises because the transverse momentum within the proton is no longer bounded, as it was in the parton model. When one integrates the quark propagator over the available phase space one obtains the integral

$$\int_{Q_0^2}^{\sim Q^2} \frac{dp_T^2}{p_T^2} \simeq \log\left(\frac{Q^2}{Q_0^2}\right) \tag{4.3}$$

where Q_0 is an arbitrary normalisation mass. Now α_s only decreases like $(\log Q^2)^{-1}$ and so (4.1) does not decrease as $Q^2 \to \infty$ as it stands. However, it is possible to restore perturbation theory by absorbing the $\log Q^2$ term into the definition of the structure function (see, for example, Field 1979). Similarly a gluon with a momentum fraction y may produce a $q\bar{q}$ pair, the resulting quark having a momentum fraction x, figure $32(b)$. The splitting function is

$$P_{g \to q}(z) = \tfrac{1}{2}[z^2 + (1 - z)^2].$$

The evolution of the quark structure function with Q^2 is thus given by

$$\frac{df_p^q(x, Q^2)}{d\log Q^2} = \frac{\alpha_s}{2\pi} \int_x^1 \frac{dy}{y}\left[P_{q \to q}\left(\frac{x}{y}\right) f_p^q(y, Q^2) + P_{g \to q}\left(\frac{x}{y}\right) f_p^g(y, Q^2)\right]. \tag{4.4}$$

In the same way we can express the Q^2 evolution of the gluon distribution within the proton as

$$\frac{df_p^g(x, Q^2)}{d\log Q^2} = \frac{\alpha_s}{2\pi} \int_x^1 \frac{dy}{y}\left[\sum_q P_{q \to g}\left(\frac{x}{y}\right) f_p^q(y, Q^2) + P_{g \to g}\left(\frac{x}{y}\right) f_p^g(y, Q^2)\right] \tag{4.5}$$

where the sum is over the different flavours of quarks. As before the 'splitting' probability functions can be calculated from the QCD Feynman rules, and are found to be

$$P_{q \to g}(z) = \frac{4}{3}\left(\frac{1+(1-z)^2}{z}\right)$$

(4.6)

$$P_{g \to g}(z) = 6\left(\frac{z}{1-z} + \frac{1-z}{z} + z(1-z)\right).$$

Although we can not determine the structure functions, $f(x, Q^2)$, from first principles, we see from (4.4) and (4.5) that, once we have measured them for one value of Q^2, we do know how they will change with Q^2. In practice these equations are not very easy to use because the gluon distributions have to be inferred indirectly (though they cancel from the evolution of the structure function difference $q - \bar{q}$), and because of the effects of higher orders of perturbation theory. The interested reader may like to consult the review by Reya (1981) for a comparison of this theory of scale breaking with the data on deep inelastic scattering. It seems that the scaling violations which occur are at least consistent with QCD.

Precisely similar considerations apply to the fragmentation functions, i.e. $D_c^C(z) \to D_c^C(z, Q^2)$, where the change in D due to the emission of partons (as in figure 35) is given by

$$\frac{dD_q^C(z, Q^2)}{d \log Q^2} = \frac{\alpha_s}{2\pi} \int_z^1 \frac{dy}{y}\left[P_{q \to q}(y) D_q^C\left(\frac{z}{y}, Q^2\right) + P_{q \to g}(y) D_g^C\left(\frac{z}{y}, Q^2\right)\right]$$

$$\frac{dD_g^C(z, Q^2)}{d \log Q^2} = \frac{\alpha_s}{2\pi} \int_z^1 \frac{dy}{y}\left[\sum_q P_{g \to q}(y) D_q^C\left(\frac{z}{y}, Q^2\right) + P_{g \to g}(y) D_g^C\left(\frac{z}{y}, Q^2\right)\right]. \quad (4.7)$$

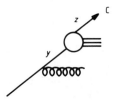

Figure 35. A scaling violation contribution to the fragmentation function D_q^C.

There are also parton emission corrections to the basic parton scattering processes of figure 31.

Clearly it is important to know if the structure and fragmentation functions, $f_A^a(x, Q^2)$ and $D_c^C(z, Q^2)$, are process-independent. Then the parton distributions determined from one process (like $ep \to eX$) could be used to help predict hadronic processes (like $pp \to \pi^0 X$), just as in the simple parton model of chapter 3. Indeed it has been shown that all divergent perturbative contributions to the processes can be summed, factored off and absorbed into universal quark and gluon distributions; and that only the Born approximation diagrams of figure 31 need be included, provided that $\alpha_s \to \alpha_s(Q^2)$. A discussion of, and references to, this powerful factorisation result can be found, for example, in Ellis and Sachrajda (1979) and Reya (1981).

To leading order the scaling violations to the hadronic process $AB \to CX$ can therefore be included in an elegant and simple way. All we have to do is to replace (3.4) by

$$E_C \frac{d\sigma}{d^3 p_C} = \sum_{abcd} \int_0^1 dx_a \int_0^1 dx_b f_A^a(x_a, Q^2) f_B^b(x_b, Q^2) \frac{1}{\pi z_c} \frac{d\sigma}{d\hat{t}}(ab \to cd) D_c^C(z_c, Q^2)$$

(4.8)

where the parton distributions are those measured in deep inelastic lepton scattering and where the Born diagrams of figure 31 are used, with $\alpha_s(Q^2)$, to calculate the hard-scattering sub-process $ab \to cd$. There remains, however, a problem as to what to take for Q^2 in α_s. For different diagrams $Q^2 = \hat{s}, \hat{t}, \hat{u}$, or some combination thereof might seem appropriate. A common practice is to employ a symmetrical combination such as $Q^2 = \hat{s}\hat{t}\hat{u}/(\hat{s}^2 + \hat{t}^2 + \hat{u}^2)$. Such substitutions make no difference to the leading log Q^2 term, but leave a significant ambiguity in the low p_T prediction.

It is found that with these changes the cross section is predicted to behave like p_T^{-6} approximately at large p_T (Field 1979). This is certainly better than the p_T^{-4} behaviour of the naive parton model, and seems fairly satisfactory for the $p_T > 6 \text{ GeV}/c$ data of figure 37, but does not account for the higher powers of p_T observed for $p_T < 6 \text{ GeV}/c$ in so many processes.

4.3 Intrinsic parton k_T

A rather serious defect of the parton model of chapter 3 is that the partons are assumed to be travelling collinearly with their parent hadron with no motion in sideways directions. In fact, this cannot be correct because the partons are confined within the radius of the hadron (typically of the order of 0.5 fm) and so by the uncertainty principle must have a momentum spread ('Fermi motion') $\Delta k \approx \hbar/(0.5 \text{ fm}) \approx 0.4 \text{ GeV}/c$ transverse to the hadron's direction. This has the consequence that the p_T of the observed hadron, C, may in part be due to the fact that partons a and b already had a net sideways motion, as in figure $36(a)$, even before scattering occurred. We denote this transverse momentum of the initial state by

$$k_T = k_{aT} + k_{bT}.$$

(4.9)

Furthermore the partons may acquire transverse momentum by parton emission before scattering. Thus in figure $36(b)$ parton a recoils from a radiated gluon before hitting b. This effect can be calculated perturbatively, at least for large k_T. It is often added to the Fermi motion, though it is not clear that these are entirely separate effects since if the gluon is reabsorbed by the other fragments of A it obviously constitutes part of the binding of a into A which creates the Fermi motion.

This intrinsic k_T can be observed rather directly in the Drell–Yan process $AB \to \mu^+ \mu^- X$. If only the lowest-order diagram, figure 26, is considered then the $\mu^+ \mu^-$ pair should have no net transverse momentum, i.e. $p_T \equiv q_{T\mu^+} + q_{T\mu^-} = 0$, but in practice $p_T \neq 0$ because quarks a and b have transverse momentum, and $p_T = k_T$. The p_T distribution is found to be approximately Gaussian $[\exp(-p_T^2/\langle p_T^2 \rangle)]$ with $\langle p_T^2 \rangle \approx 1.9 \text{ (GeV}/c)^2$, roughly independent of s and Q^2, corresponding to an effective $\langle k_T \rangle \approx 0.85 \text{ GeV}/c$ per parton.

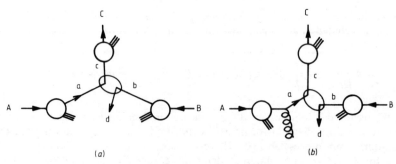

Figure 36. A sketch of (a) the intrinsic k_T due to sideways motion of partons a and b and (b) the 'effective' k_T due to bremsstrahlung, of partons before the hard scattering sub-process ab → cd. The effect of the bremsstrahlung contributions can be calculated using perturbative QCD, but only at large k_T; diagram (b) shows a lowest-order α_s contribution to the k_T of parton a.

The inclusion of this effect in large p_T hadron production is not straightforward. The difficulty stems from the fact that the tail of the Gaussian k_T distribution permits particle C to be produced with $p_T = k_T$, i.e. its transverse momentum stems entirely from the intrinsic k_T of the partons, so $\bar{t} = 0$ in the parton scattering process ab → cd. However, the Born approximations of table 3 contain an IR divergence for $\bar{t} = 0$ and so the cross section for C production is predicted to be infinite! This nonsensical conclusion stems from the fact that we are still trying to use table 3 for small $\bar{t}$ where $\alpha_s(\bar{t})$ is very large, and the perturbation expansion in α_s, of which these results are just the first term, breaks down.

It is possible to circumvent this difficulty by some such device as replacing $\bar{t} \to \bar{t} - t_0(t_0 > 0)$ in table 3, or by setting $d\sigma/d\bar{t} = 0$ for $|\bar{t}| < t_0$ in (4.8), but then the final result is quite sensitive to the method adopted and the choice of t_0, which becomes an arbitrary free parameter. Moreover, the interacting partons are off their mass shell by an amount which depends on k_T, and if the correct off-shell kinematics are used in (4.8) this intrinsic transverse momentum does not make so much difference.

It has been found that by including all the above corrections to the naive parton model of chapter 3 the data on pp → π^0X (which has the greatest range of p_T of any process so far measured) can be fitted successfully. An example is shown in figure 37. Very roughly the p_T^{-n} behaviour is changed from $n = 4$ in the naive parton model to 5 with $\alpha_s(Q^2)$, $n \approx 6$ when f, D are functions of Q^2, and $n \approx 8$ for $2 < p_T < 6 \, \text{GeV}/c$ when intrinsic k_T is included. However, quite apart from the reservations one must feel as to whether the intrinsic k_T has been incorporated correctly, explaining the data in this way leads one to expect a similar behaviour, with only minor changes due to different f and D, for all large p_T cross sections. At least for moderate $p_T(2 < p_T < 6 \, \text{GeV}/c)$ this is not borne out by experiment, so some further corrections are clearly necessary.

4.4 Other contributions

The corrections which we have included so far should be sufficient in the leading log approximation, i.e. for log $(p_T^2/\Lambda^2) \gg 1$ where Λ is the QCD mass scale parameter ($\Lambda \leqslant 1 \, \text{GeV}$). However, there are many other corrections of the form $p_T^{-m}(m > 4)$

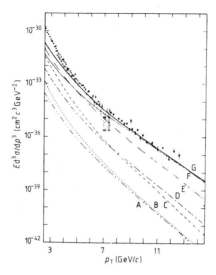

Figure 37. A fit to $pp \to \pi^0 X$ data at large p_T by Owens *et al* (1978) showing the individual sub-process contributions. The agreement at small p_T can be somewhat improved by including intrinsic k_T effects. A, $q\bar{q} \to gg$; B, $gg \to q\bar{q}$; C, $gg \to gg$; D, $q\bar{q} \to q\bar{q}$; E, $gq \to gq$; F, $qq \to qq$; G, total.

which may be important at moderate values of p_T. These arise if more than the minimal number of partons are involved in the scattering process.

For example, in inclusive meson production $AB \to M'X$, instead of the basic $qq \to qq$ process and similar processes of figure 31, we might have the sub-process $qM \to qM'$ where M is a $q\bar{q}$ cluster of fixed mass and M' is the detected 'trigger' particle (see figure 38). Comparing the two competing types of sub-process we have

$$\frac{d\sigma/d\bar{t}\,(qM \to qM')}{d\sigma/d\bar{t}\,(qq \to qq)} \sim F_M^2\,(p_T^2) \sim p_T^{-4} \qquad (4.10)$$

where F_M is the meson transition form factor. Obviously the $qM \to qM'$ contribution is not as important as $qq \to qq$ at large p_T, but it may be quite significant at moderate values of p_T. There is, after all, a large probability of finding the meson M in the incoming hadron (recall the meson 'cloud' of the proton) and if M' of the sub-process is the 'trigger' particle there is no further diminution of the cross section due to a final $q \to M'$ fragmentation. For $qq \to qq$ the outgoing quark gives only a part of its

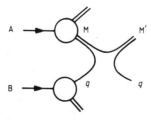

Figure 38. The $qM \to qM'$ contribution to $AB \to M'X$ drawn as in the constituent interchange model (CIM).

momentum to the trigger particle M', so the sub-process occurs at an effectively higher p_T where the cross section is much smaller. We speak of 'trigger bias' since it is only when there is a sufficiently high-energy particle that the apparatus is triggered to record the jet, but because of the fragmentation most quark jets do not produce high-energy particles. The single particle production cross section is thus much smaller than the cross section for producing a quark jet (see (5.16) below).

Basic processes of the type $qM \to qM'$ often involve the exchange of quarks and are called 'constituent interchange mechanisms', or CIM for short (Sivers *et al* 1976) as in figure 38. According to the dimensional counting rules of § 2.3 each CIM diagram gives a $(p_T^2)^{2-n}$ contribution to the inclusive cross section, where n is the number of partons taking part in the sub-process. Thus for $qM \to qM'$ we have $n = 6$ leading to a p_T^{-8} behaviour, as expected from (4.10). The process $\bar{q}q \to M\bar{M'}$ also has $n = 6$, whereas $MM \to MM'$ has $n = 8$ and so on. Similarly we expect the structure functions will have corrections of the form $f_A'^a(x)(Q^2)^{-m}$ with $m = 2, 4, \ldots$, due to internal gluon exchanges, etc, as in figure 39. For technical reasons these are called 'higher

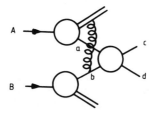

Figure 39. A higher twist diagram which involves an interaction with a 'spectator' parton.

twist' effects (see Buras 1980) and though clearly irrelevant for large Q^2 the fact that their magnitude is unknown makes the precise determination of the logarithmic scaling violations of the leading term ($n = 4$) difficult.

One thus ends up with a sequence of terms:

$$E_C \frac{d^3\sigma}{d^3 p_C}(AB \to CX) = F_1(\theta_{cm}, x_T)p_T^{-4} + F_2(\theta_{cm}, x_T)p_T^{-6} + \cdots \qquad (4.11)$$

each of which should enjoy logarithmic scaling corrections. Some fairly satisfactory fits to the data along these lines have been made (Jones and Gunion 1979), but the presence of arbitrary normalisation parameters makes their significance hard to assess. In $pp \to \pi^0 X$ the p_T^{-4} parton scattering mechanism does not seem to dominate until $p_T > 10 \, GeV/c$, and instead for $2 < p_T < 6 \, GeV/c$ the $p_T^{-8} \, qM \to qM$ contribution is the most significant. This is also true in $pp \to \pi^\pm X$, whereas in $pp \to pX$ the behaviour is p_T^{-12} as expected for $qB \to qp$.

These CIM processes involve the exchange of quarks and hence of flavour quantum numbers. This can be tested by comparing the quantum numbers of the trigger particle, C, with those of the particles X which have the opposite sign for p_T and which presumably result from the fragmentation of d (Brodsky 1979a, b). With gluon exchange there will be no quantum number correlations between the two sides whereas with quark exchange the opposite-side particles are likely to have the opposite sign of charge (and other additive quantum numbers) to C. Some such correlation is in fact observed at moderate p_T (see figure 40) and argues in favour of CIM contributions.

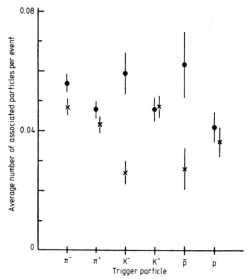

Figure 40. Charge correlations for various types of trigger taken from Albrow *et al* (1978). The dots and crosses are the average number of fast positive and negative particles, respectively, on the side away from a 90° trigger. $3 < p_{\text{trig}} < 4.5 \, \text{GeV}/c$, $p_{\text{T}} > 1.5 \, \text{GeV}/c$, $|y| < 1$, $|\varphi| < 30°$.

On the other hand, it is found that the cross sections for $\pi^- \text{p} \rightarrow \pi^- \text{X}$ and $\pi^- \text{p} \rightarrow \pi^+ \text{X}$ are almost equal for $2 < p_{\text{T}} < 6 \, \text{GeV}/c$, as one would expect for the flavour-blind gluon exchange, but quite contrary to CIM which predicts the π^- should predominate due to hard scattering of the incident π^-.

The intermediate p_{T} region (2–6 GeV/c) is thus rather confusing, and neither parton scattering nor CIM seems able to explain all aspects of the data, but at larger p_{T} the parton model with QCD corrections does seem to be in quite good agreement with the admittedly modest quantity of data available to date.

5

JETS

5.1 Introduction

After a hard scattering process such as figure 24 there will be partons flying apart
with a large relative momentum, but carrying a colour charge, which we believe to
be a confined quantity. The partons must thus dispose of their momentum by radiating
colourless hadrons. It is this process which we have parametrised by the fragmentation
functions $D_c^C(z, Q^2)$ in the preceding sections. We now want to look at the way this
actually happens in a bit more detail (see, for example, Dokshitzer *et al* 1980, Webber
1982).

It is convenient to consider first $e^+e^- \rightarrow$ hadrons since in this process there are no
coloured partons in the initial state to complicate matters (see, for example, Wiik and
Wolf 1979, Wolf 1980). The parton model description is shown in figure 41(*a*) in

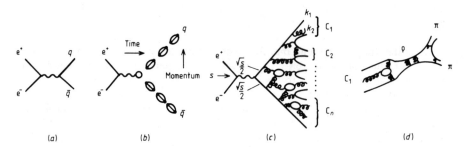

Figure 41. The process $e^+e^- \rightarrow$ hadrons.

which initially there is a quark and an antiquark, each carrying colour charge and
trying to leave the interaction region. The colour lines of force thus get stretched and
so the force between the quarks increases, rather as if they were attached to the two
ends of an elastic band. One might at first suppose that this $q\bar{q}$ pair would therefore
be compelled to remain as a single $(q\bar{q})$ hadron, at rest in the e^+e^- centre-of-mass
system but with a very large internal energy (i.e. $\sqrt{s} =$ hadron mass). However, as
we noted in chapter 1, it is energetically more favourable for the 'string' to break by
the formation of additional $q\bar{q}$ pairs (i.e. colour polarisation of the vacuum occurs)
leaving a collection of colourless states with only rather short lines of force (~ 1 fm)
which are, of course, low-mass hadrons, mainly pions.

The way in which these hadrons may develop with time is shown in figure 41(*b*)
and a typical Feynman-like diagram representing such an event in figure 41(*c*) (Konishi
et al 1979). At the photon vertex there is a q and a $\bar{q}$, each with momentum $\approx \sqrt{s}/2$,
but confined because they carry colour and so the force between them will increase
as they try to separate. But each successive branching (gluon emission, $q\bar{q}$ creation,
etc) not only reduces the momentum of the partons but also helps to screen the colour

charges. Eventually clusters can be found with $k_C^2 = (k_1 + k_2 + \ldots)^2 = m_C^2$ in which $m_C^2 = m_h^2$, the mass of some hadron. If such a cluster has zero net colour it is likely to combine to form this hadron (since $\alpha_s(m_C^2)$ is large), which is then free to leave the interaction region (figure 41(d)). In this way we expect two back-to-back jets of hadrons to occur (figure 42), sharing between them the momentum of the initial q or $\bar{q}$, and each having only a small momentum component transverse to the quark's direction of motion, $p_T \lesssim \Lambda$ (Amati and Veneziano 1979).

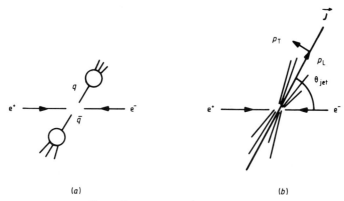

(a) (b)

Figure 42. The process $e^+e^- \rightarrow$ two jets.

As we shall see later, it is also possible for one of the quarks to emit a 'hard' gluon (which has a large transverse momentum component, $k_T \gg \Lambda$), in which case there will be a three-jet event, as in figure 43.

In the initial stages of this evolution all the virtual parton masses are large, $k_i^2 \gg \Lambda^2$, and so $\alpha_s(k_i^2)$ is small and the diagrams can be evaluated by the usual rules of Feynman perturbation theory (Ellis *et al* 1976b). In fact, the probability of a particular branching process occurring is given by (4.1) and (4.6) and hence a perturbative 'jet calculus' can be derived (Konishi *et al* 1979). But in the final stages where the partons are combining into hadrons (figure 41(d)) α_s is not small and so they are not amenable to QCD calculations so far. Despite this we can reasonably expect that the sum of the momenta of the hadrons in a jet will correspond (to an accuracy $\simeq \Lambda$) to the momentum of the parent parton.

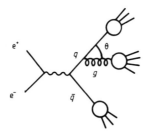

Figure 43. The process $e^+e^- \rightarrow$ three jets.

5.2   Jets in $e^+e^- \rightarrow$ hadrons

To test whether such jets are in fact occurring experimentally it is useful to define a measure of the degree of 'jettyness' exhibited by the hadrons produced in an event (Wiik and Wolf 1979). In the centre-of-mass system the q and $\bar{q}$ are initially moving with equal and opposite momentum along some jet axis J (figure 42). We expect the hadrons in the jets to be travelling more or less in this direction too, and so to find the direction of J we look for the axis such that the magnitudes of the particles' momentum components, p_T, transverse to J are minimised, while their longitudinal components, p_L, are maximised. Two such jet measures are (Bjorken and Brodsky 1970, Farhi 1977)

$$\text{sphericity} \qquad S \equiv \min_{J} \left(\frac{3}{2} \frac{\Sigma_i p_{T_i}^2}{\Sigma_i p_i^2} \right) \tag{5.1}$$

$$\text{thrust} \qquad T \equiv \max_{J} \left(\frac{\Sigma_i |p_{L_i}|}{\Sigma_i p_i} \right) \tag{5.2}$$

where the sum runs over all the particles i in the event, and the extrema are chosen by varying the direction of J. If there were no jets and all momentum directions were equally probable, then $S = 1$ and $T = \frac{1}{2}$, irrespective of the choice of J. For perfect jets all the hadrons would be travelling along a given axis and so the extrema would be $S = 0$ and $T = 1$ when J is along that axis.

Figure 44 illustrates the way in which the average sphericity and thrust of $e^+e^- \rightarrow$ hadrons events vary as a function of the energy and it will be seen that $\langle S \rangle$ and $1 - \langle T \rangle$ do indeed become quite small at high energies. They do not vanish, however, indicating that some hadrons always have a finite p_T.

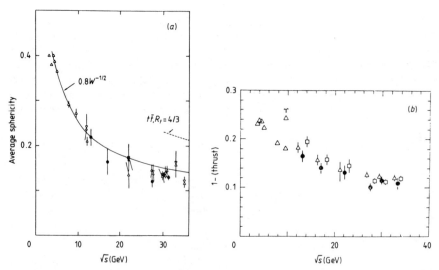

Figure 44. The average sphericity (a) and $1 -$ thrust (b) as a function of the e^+e^- CM energy (Wolf 1980). Ξ JADE, $\triangle$ PLUTO, $\bullet$ TASSO, $\square$ MARK-J.

In part this is presumably due to the confinement mechanism. We have noted in chapter 4 that partons which are confined within hadrons have a Fermi momentum component perpendicular to the hadron's direction of motion, with $\langle k_T^2 \rangle \approx 0.3\,\text{GeV}^2$. We must thus expect that any hadrons radiated from a parton will have a complementary $\langle p_T^2 \rangle \approx 0.3\,\text{GeV}^2 (\approx \Lambda^2)$ and this is borne out by figure 45. The jet direction thus becomes better defined as the jet energy ($\frac{1}{2}\sqrt{s}$), and hence the average p_L, increases, and the jet opening angle decreases with energy roughly like

$$\theta \sim \frac{\langle p_T \rangle}{\langle p_L \rangle} \sim \frac{1}{\sqrt{s}}. \tag{5.3}$$

However, it will be seen that $\langle p_T^2 \rangle$ increases slowly with energy. This behaviour is also expected in QCD because we must also consider diagrams such as figure 43, for

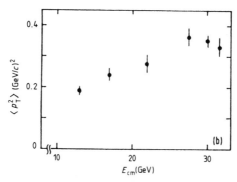

Figure 45. The average transverse momentum squared of hadrons with respect to the jet axis (Wolf 1980).

example, in which one of the quarks radiates a gluon. If we denote the energy fractions carried by the three final-state partons as

$$x_i = 2E_i / \sqrt{s} \qquad i = q, \bar{q}, g \tag{5.4}$$

then for massless partons the differential cross section for the process $e^+e^- \to q\bar{q}g$ is

$$\frac{d^2\sigma}{dx_q\,dx_{\bar{q}}} = \frac{2\alpha_s}{3\pi}\,\sigma(e^+e^- \to q\bar{q})\,\frac{x_q^2 + x_{\bar{q}}^2}{(1-x_q)(1-x_{\bar{q}})}. \tag{5.5}$$

This expression introduces one of the principal problems of using QCD perturbation theory, for it diverges as x_q or $x_{\bar{q}} \to 1$. Now by four-momentum conservation

$$s(1-x_{\bar{q}}) = (p_q + p_g)^2 = 2p_q \cdot p_g = 2E_q E_g(1 - \cos\theta) \tag{5.6}$$

where θ is the angle between the outgoing quark and the gluon. So $x_{\bar{q}} \to 1$ if either $E_g \to 0$, i.e. if the gluon carries no energy, or if $\theta \to 0$, i.e. if the gluon travels along the direction of its parent quark. These are known as the infrared (IR) and collinear divergences, respectively, and plague any theory containing massless particles (including QED) (Dokshitzer *et al* 1980, Webber 1982). Indeed we have already met the IR divergence in §§ 4.2 and 4.3.

For small E_g and θ (5.5) gives

$$\frac{d^2\sigma}{dE_g\,d\theta} \sim \frac{\alpha_s}{E_g\theta}\,\sigma(e^+e^- \to q\bar{q}) \tag{5.7}$$

and so the average transverse momentum of the gluon is

$$\langle k_T \rangle \sim \frac{\alpha_s \sigma}{\sigma} \int \frac{E_g \sin\theta}{E_g \theta} \, dE_g \, d\theta \sim \alpha_s \sqrt{s}. \tag{5.8}$$

If this is reflected in the transverse momentum of the hadrons then (5.3) suggests that the jet opening angle will decrease only logarithmically at very high s. This is presumably the origin of the observed slow increase of $\langle p_T^2 \rangle$ with s in figure 45.

If E_g and θ are not small we can expect to see a third (gluon) jet, in addition to the two quark jets, and such events have indeed been observed in high-energy e^+e^- annihilation experiments. Since in (5.5) the relative probability of $q\bar{q}g/q\bar{q}$ is of the order of α_s (see Wiik and Wolf 1979), the magnitude of the three-jet cross section relative to that for two jets can be used to determine that $\alpha_s \approx 0.18$ for $\sqrt{s} \approx 30$ GeV (Wiik 1980).

The divergence in (5.5) will obviously give a divergent contribution to the total cross section for $e^+e^- \to$ hadrons. However, we know that real massless gluons are not seen, and this divergence can be 'regularised' away by giving the gluon a finite mass, m_g (see, for example, Field 1979a,b), in which case the $x_{\bar{q}} \to 1$ region of the integration of (5.5) over the x gives

$$\sigma(e^+e^- \to q\bar{q}g) \simeq \sigma(e^+e^- \to q\bar{q}) \frac{2\alpha_s}{3\pi} \log\left(\frac{m_g^2}{s}\right). \tag{5.9}$$

This dependence on the gluon mass looks unpleasant, but in fact it can be shown that this term is precisely cancelled by the corresponding $O(\alpha_s)$ correction to $\sigma(e^+e^- \to q\bar{q})$ shown in figure 46(c), and that the sum of the diagrams in figure 46 leaves us with the $O(\alpha_s)$ correction:

$$\sigma(e^+e^- \to \text{hadrons}) = \sigma(e^+e^- \to \mu^+\mu^-) \sum_q e_q^2 \left(1 + \frac{\alpha_s}{\pi}\right) \tag{5.10}$$

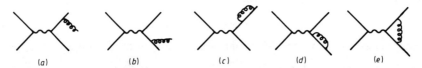

Figure 46. Lowest-order contributions to $e^+e^- \to q\bar{q}g$ (a), (b) and corrections to $e^+e^- \to q\bar{q}$ (c), (d), (e).

where $\sigma(e^+e^- \to \mu^+\mu^-)$ is the cross section (2.15), and the final answer, (5.10), is independent of m_g despite (5.9).

The total hadron cross section is thus IR-safe (certainly to first order in α_s and presumably to all orders), i.e. it has no singularities as $m_g \to 0$. This is because the total cross section is independent of the number of partons produced: it does not make any difference whether we regard the gluon in figure 43 as part of the quark jet, or as providing a separate jet. Similarly a quantity such as thrust, which depends linearly on the momenta, and hence by momentum conservation is independent of the number of partons in the final state, is IR-safe and can be calculated in perturbation theory. To first order (De Rujula et al 1978)

$$1 - \langle T \rangle = 1.57 \frac{2}{3} \frac{\alpha_s}{\pi}. \tag{5.11}$$

But the sphericity, being quadratic in the momenta, is not IR-safe, and cannot be predicted by perturbative QCD. The number of partons produced in an event is not experimentally observable since whenever off mass-shell partons find themselves in clusters having an invariant mass of order Λ, $\alpha_s \to 1$ and hadronisation occurs as described above.

Another example of how to overcome these IR problems has been given by Sterman and Weinberg (1977) (see also Webber 1982). In the naive parton model of figure 41(a) one expects that the angular distribution of the quark jet axis J in $e^+e^- \to q\bar{q}$ will be the same as that of the μ distribution in $e^+e^- \to \mu^+\mu^-$, viz

$$\frac{d\sigma}{d\Omega} = \frac{\alpha^2}{4s}(1 + \cos^2\theta) \tag{5.12}$$

where θ is the angle between the direction of motion of the muons and the direction of the electron beam in the centre-of-mass system. However, if one tries to calculate the corrections caused by higher-order processes such as $e^+e^- \to q\bar{q}g$ one again runs into divergence problems due to the massless gluons. Instead Sterman and Weinberg define an event as a two-jet event if all but a fraction $\epsilon(\ll 1)$ of the total energy of the event falls within two back-to-back cones of opening angle δ. They then use perturbative QCD to calculate the cross section in the form

$$\frac{d\sigma}{d\Omega_{\text{jet}}}(\epsilon, \delta) = \frac{d\sigma}{d\Omega_0}\sum_q e_q^2 f(\delta, \epsilon) \tag{5.13}$$

where $d\sigma/d\Omega_0$ is the QED cross section of (5.12) which has to be modified by the charges of the produced quarks, and f contains QCD corrections which in first order in α_s include terms such as $\alpha_s \log(1/\delta)\log(1/\epsilon)$ which diverge if we try to define the jets too closely by making δ or ϵ small.

Very similar jets to those in $e^+e^- \to$ hadrons are also seen in the current fragmentation region of deep inelastic scattering, such as $ep \to eX$, where the struck quark fragments seemingly independently of the spectator partons (see Di Lella 1979, Darriulat 1980).

5.3   Jets in hadron scattering

As we noted in chapter 2, large p_T hadron scattering processes are predominantly four-jet events, as in figure 17. Two of the jets contain fragments of the incoming hadrons and hence continue more or less in the forward direction, while the two wide-angle back-to-back jets, stemming from the hard scattered partons, contain particles with large p_T. Thus, accompanying the large p_T particle which has triggered the detectors to tell us that a large p_T event has occurred, there are other particles, travelling in essentially the same direction, forming a 'same-side' jet, and in the opposite hemisphere an 'away-side' jet. Figure 47 shows that large p_T particles whose rapidities are closely correlated with that of the trigger particle are found on both the same and the opposite sides. The longitudinal jets appear to be the same as those observed in ordinary hadron scattering when no large p_T particle is produced. These will be the subject of the next chapter:

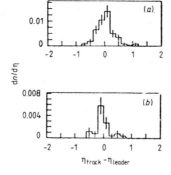

Figure 47. The rapidity correlation of the same- and away-side jets. The data are the charged particles associated with a π^0 with $p_T \geq 7$ GeV/c that is produced in the opposite hemisphere. The rapidity ($\eta = y$) is referred to that of the 'hardest' away particle (the leader), and we see that the collimation increases with the p_T of the leader. (*a*) $3 < p_T < 4$, (*b*) $4 < p_T < 5$ GeV/c.

This type of triggering on a single high p_T particle, though very convenient experimentally, greatly reduces the probability of observing jets because the fragmentation functions depress the likelihood of a single particle carrying a significant fraction of the jet's momentum (Jacob and Landshoff 1978).

To illustrate this suppose that the cross section for producing a parton jet with large P_T behaves as

$$\frac{d\sigma(\text{jet})}{dP_T} = \frac{A}{P_T^N} \tag{5.14}$$

and that the jet then fragments into hadrons carrying a longitudinal fraction z of the jet's momentum with probability $D(z)$, then

$$\frac{d\sigma(\text{particle})}{dp_T} = \int_0^1 dz \int dP_T \frac{A}{P_T^N} D(z)\delta(p_T - zP_T)$$

$$= \frac{A}{p_T^N} \int_0^1 D(z) z^{N-1} dz. \tag{5.15}$$

So if we take $D(z) = (1-z)^m/(m+1)z$, corresponding to the form (3.26) with the normalisation satisfying the momentum sum rule (3.22), then

$$\frac{d\sigma(\text{particle})/dp_T}{d\sigma(\text{jet})/dP_T}\bigg|_{P_T = p_T} = \frac{(N-2)!\, m!}{(m+1)(N+m-1)!}. \tag{5.16}$$

Thus for π production from a quark jet with $m = 2n_s - 1 = 1$ and $N = 8$, as observed, the ratio is 1/112. So single-particle cross sections should be suppressed with respect to jet cross sections by at least two orders of magnitude (three with larger m, N). The two types of cross section should show the same dependence on p_T, however. Both these expectations are borne out by figure 48, though of course much depends on how the jet cross section is defined experimentally, i.e. which particles are included in the jet and which, because they are moving too slowly or at too large an angle to the jet axis, are excluded. Nowadays jet cross sections can be obtained using calorimeters which trigger whenever sufficient total hadron energy is deposited at a

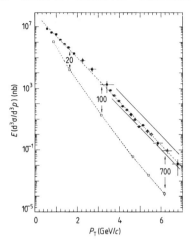

Figure 48. The inclusive jet production cross section (■) compared to inclusive single π production (□) taken from the review by Darriulat (1980).

large angle to the beam direction, and we shall discuss some of the recent results in chapter 11. Of course it is still not possible to be sure that the calorimeter has caught all the jet particles, and no others, but once it is possible to produce jets of sufficiently high energy they stand out very clearly indeed.

The best way to remove any such bias is to look at the away-side jet opposite to a high p_T trigger. Since the quark fragmentation functions are suppressed at large z less than those describing gluon fragmentation (see (3.25) and figure 30) it is more likely that the large p_T trigger particle is from a quark jet, but the opposite side may quite often be a gluon jet. If there were great differences between quark and gluon jets we might expect that these away-side jets in large p_T hadron scattering would look somewhat different from the jets in $e^+e^- \rightarrow$ hadrons and DIS. In fact, the data show that the average multiplicity (figure 49), the shape of the fragmentation functions (figure 50), and the average transverse momentum of the particles with respect to the jet axis, are all consistent in the three types of process. The modest accuracy of the data, and the uncertainty as to which particles should be included in the jets, obscure any differences there may be between quark and gluon jets. There should also be higher-order, five and more jet events when the quarks radiate hard gluons, etc, but as yet these have not been clearly distinguished.

A very interesting problem is the extent to which the flavour quantum numbers of the particles produced in a jet reflect the flavour of the parent parton. If we adopt a simple model such as figure 51(a) in which all the particles produced above the broken line are regarded as having large p_T, and hence are included in the quark jet, while those below it are slow particles and not part of the jet, then the average charge of the jet will be (Brodsky and Weiss 1977)

$$\langle Q_{\text{jet}} \rangle = Q_q + \langle Q_{\bar{q}} \rangle \qquad (5.17)$$

where Q_q is the charge of the quark and the second term on the right-hand side represents the average charge of the $\bar{q}$ line, which must also be cut by the broken line if only colour neutral particles are to appear in the jet final state.

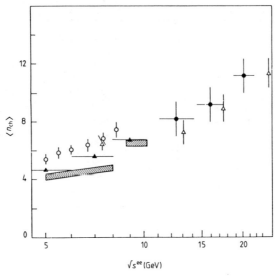

Figure 49. The mean charged multiplicity of hadron jets measured in large p_T hadronic processes ($\bigcirc$), in νp interactions ($\blacktriangle$), and in e^+e^- annihilations (shaded zones and $\triangle$), taken from Darriulat (1980). All data are referred to the equivalent e^+e^- energy.

If we suppose that $\bar{u}$ and $\bar{d}$ antiquarks can be created with equal probability (and similarly for the heavier pairs ($\bar{c}, \bar{s}$) and ($\bar{t}, \bar{b}$)) then

$$\langle Q_{\bar{q}} \rangle = \tfrac{1}{2}(Q_{\bar{u}} + Q_{\bar{d}}) = \tfrac{1}{2}(-\tfrac{2}{3} + \tfrac{1}{3}) = -\tfrac{1}{6}. \tag{5.18}$$

Alternatively, if we suppose that only those antiquarks with masses less than Λ, the QCD mass scale, are easily produced (i.e. $\bar{u}$, $\bar{d}$ and $\bar{s}$) while the heavier $\bar{c}$, $\bar{b}$, etc,

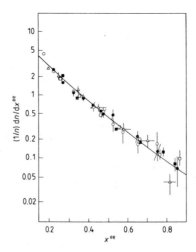

Figure 50. The jet fragmentation functions measured in pp collisions ($\bigcirc\bullet\square\blacksquare$), in νp interactions ($\triangle$) and in e^+e^- annihilations (——), taken from Darriulat (1980).

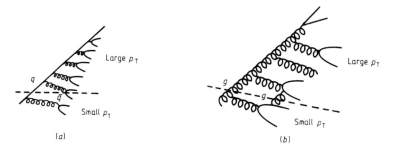

Figure 51. Colour flow in (a) quark and (b) gluon jets.

appear only very infrequently, one might expect that instead

$$\langle Q_{\bar{q}}\rangle = \tfrac{1}{3}(Q_{\bar{u}} + Q_{\bar{d}} + Q_{\bar{s}}) = \tfrac{1}{3}(-\tfrac{2}{3} + \tfrac{1}{3} + \tfrac{1}{3}) = 0. \tag{5.19}$$

The truth probably lies somewhere between (5.18) and (5.19) in which case we find

$$\langle Q_{u,\text{jet}}\rangle = \tfrac{2}{3} + (-\tfrac{1}{6} \text{ to } 0) = \tfrac{1}{2} \text{ to } \tfrac{2}{3} \qquad \text{(and for } c, t \text{ jets)}$$

$$\langle Q_{d,\text{jet}}\rangle = -\tfrac{1}{3} + (-\tfrac{1}{6} \text{ to } 0) = -\tfrac{1}{2} \text{ to } -\tfrac{1}{3} \qquad \text{(and for } s, b \text{ jets).} \tag{5.20}$$

These conclusions are best tested in deep inelastic ν scattering which (neglecting sea quarks and the Cabibbo weak mixing angles) selects almost a unique quark flavour. Thus in $\nu p \to \mu^- X$ ($\approx u$ jet only) it is found that $\langle Q_{\text{jet}}\rangle = 0.61 \pm 0.09$, while in $\bar{\nu} p \to \mu^+ X$ ($\approx d$ jet only) $\langle Q_{\text{jet}}\rangle = -0.15 \pm 0.21$, in good agreement with (5.20) (Berge *et al* 1980).

For gluon jets (figure 51(b)) the cutoff line must cross two gluon lines to ensure colour neutrality and a similar analysis suggests that $\langle Q_{g,\text{jet}}\rangle \approx 0$, but this is much harder to test experimentally.

The above arguments are readily extended to other conserved flavour quantum numbers, such as strangeness or charm. If it is the case that hadron production proceeds mainly by ordered $q\bar{q}$ creation as in figure 51 then one would expect an anticorrelation between the flavours of adjacent particles in the jet. Such an anti-correlation between the charges of the leading and next-leading particles in jets has in fact been observed (Darriulat 1980).

The gluon exchange mechanism of figure 31 suggests that there should be no correlation between the flavours of particles in opposite-side jets as we noted in chapter 4. Figure 40 shows the average numbers of + and − charge particles in the away-side jets opposite to various large p_T triggers. A significant anticorrelation of the charges is evident, though not in all experiments. If true this suggests that quark exchange (as in the CIM model, for example) is making a significant contribution to the cross section (Brodsky 1979a, b).

Obviously, much remains to be understood about the parton fragmentation process and the properties of jets in hadron collisions. So far all the evidence points to a strong similarity between these jets and those found in e^+e^- annihilation and deep inelastic scattering experiments. Much more can be learned from the really high-energy hadron colliding-beam facilities such as the CERN p$\bar{\text{p}}$ Collider where jets with really large $p_T (\lesssim 150 \text{ GeV}/c)$ are produced at a significant rate, and jet identification is less of a problem. We shall discuss these in chapter 11.

6

PARTON IDEAS FOR LOW p_T HADRONIC INTERACTIONS

The success of the parton approach in hard scattering processes prompts the question of its relevance to hadronic processes at small momentum transfer. It might be thought that phenomena in the low p_T regime would be too complicated to be explained in terms of partons. There appears to be no large Q^2 in the problem to justify the use of perturbative QCD. However, in the last few years there has been much activity, particularly in trying to explain the longitudinal momentum (or x) distribution of fast hadrons at low p_T in terms of the quark structure and fragmentation functions. There is a variety of different dynamical models, among which are the quark chain model (Capella *et al* 1979), quark fragmentation model (Andersson *et al* 1977), quark recombination model (Das and Hwa 1977, Hwa 1980) and a dual topological unitarisation approach (Cohen-Tannoudji *et al* 1980). The confusion is increased in that many of these models seem to have very different bases and yet all claim success in describing the data. A review of the parton approaches to low p_T physics has been given by Fiałkowski and Kittel (1983).

As long ago as 1972 a simple additive quark model was used to estimate the hadron yields in the central region of hadron–hadron collisions (Anisovich and Shekhter 1973). For a meson–nucleon collision the idea is shown in figure 52. The

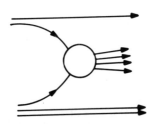

Figure 52.

interaction, shown as a blob, involves a single q (or $\bar{q}$) from each hadron. At high energies many quarks and antiquarks are produced, and the hadron production has no memory of the initial state. In other words, the number of $q\bar{q}$ pairs in the hadron sea is so large in the small x region that the influence of the initial quantum numbers is negligible. The distribution of particles in the central region should therefore be independent of the incident hadrons.

To make a hadron an outgoing quark at small x must pick up at least one neighbouring quark. At high energies there should be a 50% chance of a q grabbing a $\bar{q}$ to make a meson; in the other half of cases a qq system is formed which must combine again, and is equally likely to form qqq (baryon) or $qq\bar{q}$, etc. The particle yields are calculated assuming the statistical equality of all possible q and $\bar{q}$ combinations. Due to the relative scarcity of observed hadrons containing strange quarks a suppression factor λ is introduced for each strange quark produced (presumably due to its greater mass). The dominant production is taken to be 0^-, 1^- mesons and $\frac{1}{2}^+$, $\frac{3}{2}^+$ baryons and antibaryons. The effect of resonance decays on the particle yields is included; in fact most of the π arise via the decay of resonances, such as the ρ. This simple model, with $\lambda \approx 0.3$, successfully explains the relative meson yields, and the universal x dependence, in the central region (see, for example, Kittel 1981). In particular, since quarks and antiquarks are assumed to be produced with uncorrelated spin projections, the number of $q\bar{q}$ pairs with total spin S is proportional to the statistical weight $2S + 1$. The predicted ratios, for example $3:1$ for K*(890):K production, are in agreement with the data.

In the fragmentation region $(x \to \pm 1)$ the situation is quite different. This is the region in which the produced particle has acquired a large fraction of the incident longitudinal momentum, and as expected the spectrum is dependent on the quantum numbers of the fragmenting particle.

6.1 Fragmentation versus recombination

To illustrate the application of parton ideas to low p_T fragmentation consider the π^+ inclusive cross section in a proton–proton collision; $pp \to \pi^+X$. A π^+ produced with a large fraction of the fragmenting proton's momentum, say $x \sim 0.8$, is expected to arise from a quark whose momentum fraction is also large. Such a quark is most likely to be a u valence quark of the initial proton moving in the same direction. We expect this fast quark to be unaffected by the hadron collision, and to be described by the same structure function, $f_p^u(x)$, found in deep inelastic lepton scattering (see § 3.2). Thus to predict $pp \to \pi^+X$ we need to understand the mechanism by which a π^+ can be formed from such a fast u quark. One possibility is that the u fragments into π^+ by the mechanism that we used for high p_T processes. But at low p_T there is the alternative possibility that the u combines with a slow $\bar{d}$ from the proton sea to form a π^+. These two mechanisms, known as fragmentation and recombination respectively, are sketched in figure 53. They will be discussed in turn.

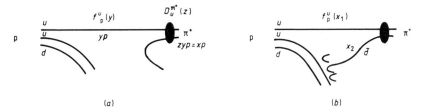

Figure 53. (a) The fragmentation and (b) the recombination mechanism for a proton fragmenting into a π^+ meson.

6.1.1 The fragmentation model

Using the fragmentation mechanism (figure 53(a)), the $pp \to \pi^+X$ inclusive cross section takes the form

$$\frac{1}{\sigma} \frac{d\sigma^{(p \to \pi^+)}}{dx} = \int_x^1 f_p^u(y) D_u^{\pi^+}(x/y) dy/y. \tag{6.1}$$

But with the known structure and fragmentation functions (see chapter 3) this yields an answer more than an order of magnitude below the data (Das and Hwa 1977). The trouble is that the momentum is lost in two stages ($x = yz$); first, the quark carries only a fraction y of the incident proton's momentum, and second, it then fragments into a π^+ together with other particles which compete for the momentum.

We can look at this deficiency using the spectator quark counting rules, (3.14), which predict

$$f_p^u(y) \sim (1-y)^3 \qquad D_u^{\pi^+}(z) \sim (1-z). \tag{6.2}$$

Inserting this behaviour in (6.1) gives

$$\frac{d\sigma^{(p \to \pi^+)}}{dx} \sim (1-x)^{?n_s-1} \sim (1-x)^5 \tag{6.3}$$

where n_s is the total number of spectator quarks. In contrast, experiments reveal a π^+ momentum behaviour $(1-x)^\beta$ with β near 3.

The various models based on quark fragmentation avoid this difficulty by a mechanism known as the quark 'held-back' effect. For example, for the meson fragmentation process, $\pi^+ p \to hX$, one valence quark is held back in the central region ($x \simeq 0$) and the other valence quark (plus the gluons and sea) fragments as if it carried all the π^+ momentum. Since there are equal probabilities for the u or $\bar{d}$ to go forward and fragment

$$\frac{1}{\sigma} \frac{d\sigma^{(\pi^+ \to h)}}{dx} = \tfrac{1}{2}[D_u^h(x) + D_{\bar{d}}^h(x)]. \tag{6.4}$$

Meson fragmentation data are found to compare well with the predictions obtained using the quark fragmentation functions determined from leptoproduction data (Andersson *et al* 1977). Indeed, the motivation for the fragmentation model is the surprising similarity between multiparticle production mechanisms in e^+e^- annihilation, in leptoproduction and in low p_T hadron–hadron interactions (see figures 49 and 50). We speak of 'jet universality'. In proton fragmentation a valence quark is held back and the remaining diquark system goes forward and fragments, and so, for example we have

$$\frac{1}{\sigma} \frac{d\sigma^{(p \to \pi^\pm)}}{dx} = \tfrac{2}{3} D_{ud}^{\pi^\pm}(x) + \tfrac{1}{3} D_{uu}^{\pi^\pm}(x). \tag{6.5}$$

Diquark fragmentation functions obtained from recent deep inelastic neutrino data appear to satisfy this equality (see, for example, Kittel 1981).

The 'held-back' effect is also operative in the quark chain approach of the Orsay group (Capella *et al* 1979, 1980). Although this fragmentation model is formulated in terms of parton concepts it is inspired by the dual topological unitarisation scheme. It assumes that the interaction separates the valence quarks of each incident proton into two coloured systems, one with the quantum numbers of a valence quark and the other those of a diquark. To neutralise these coloured systems, two multiparticle chains are formed stretching from one proton to the other, as in figure 54(a). It is argued that the valence quarks are held back in the central region and that the diquarks carry most of the proton momenta. Squaring the diagram and summing over inter-

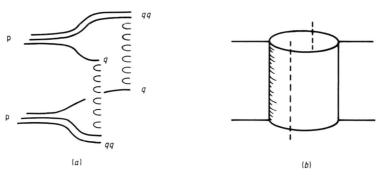

Figure 54. The cylinder topology of diagram (*b*) is formed by the 'back-to-back' product of amplitudes of diagram (*a*).

mediate states to obtain the cross section, as required by unitarity, gives the cylinder topology of the Pomeron in the dual model (figure 54(*b*)). This provides some justification for the two-chain colour separation, at least in the central region.

Attempts have been made to put this held-back mechanism on a firmer theoretical basis by invoking the concept of stretching the colour flux tube (Andersson *et al* 1980), or through the dual model (Cohen–Tannoudji *et al* 1980).

It is clearly important to compare the rapidity distribution of the particles produced in pp interactions with that in e^+e^- annihilation which arises from a single multiparticle chain (see figure 55). Comparing data at 30 GeV, we find that the average charged-particle multiplicities and the central plateau heights satisfy

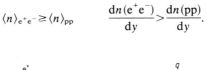

$$\langle n \rangle_{e^+e^-} \gtrsim \langle n \rangle_{pp} \qquad \frac{dn(e^+e^-)}{dy} > \frac{dn(pp)}{dy}.$$

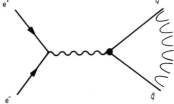

Figure 55. Schematic diagram of the process $e^+e^- \rightarrow$ hadrons.

At first sight this appears to rule out the two-chain model (figure 54) from which naively we would predict $\langle n \rangle_{pp} \simeq 2 \langle n \rangle_{e^+e^-}$. However, the 'held-back' effect may restore agreement with the data. First, the plateaux from the two multiparticle chains are each moved away from $y = 0$ so that they only partially overlap in rapidity. Secondly, only the sub-energy determines the plateau height rather than the full energy as in e^+e^-. Consistency with existing data is possible (Capella *et al* 1980) and it will be interesting to see if the two-chain model correctly predicts the development of the plateaux with increasing energy.

6.1.2 *The recombination model*

The recombination model (figure 53(b)) appears to be an entirely different approach to low p_T fragmentation. The basic idea is that the fast valence quark recombines with a slow ($x_2 \simeq 0$) antiquark from the sea. Now $x = x_1 + x_2 \simeq x_1$, and so for the example pp $\rightarrow \pi^+$X

$$\frac{d\sigma^{(p \rightarrow \pi^+)}}{dx} \sim f_p^u(x) \sim (1-x)^3 \tag{6.6}$$

in reasonable agreement with experiment, and in contrast to the prediction of (6.3). The original idea goes back several years (Goldberg 1972, Pokorski and Van Hove 1975), but the present impetus originates from the observation by Ochs (1977) that the ratio of π^+ to π^- inclusive production by protons satisfies

$$\frac{d\sigma^{(p \rightarrow \pi^+)}/dx}{d\sigma^{(p \rightarrow \pi^-)}/dx} \simeq \frac{f_p^u(x)}{f_p^d(x)}$$

with the quark structure functions determined from inelastic lepton scattering data (see § 3.2). This follows from (6.6) and the analogous relation for p $\rightarrow \pi^-$.

In detailed applications of the model, allowance is made of the momentum fraction x_2 carried by the sea quark (see figure 53(b)). A knowledge of the joint $q - \bar{q}$ momentum probability distribution and the $q - \bar{q}$ recombination function is required. There have been several applications to determine meson structure functions among which are Das and Hwa (1977), Duke and Taylor (1978), Hwa and Roberts (1979) and Aitkenhead *et al* (1980).

A recent theoretical development is the valon recombination model of Hwa (1980). In this approach hadron fragmentation proceeds as follows:

initial hadron $\rightarrow$ valons $\rightarrow$ partons $\rightarrow$ valons $\rightarrow$ final hadron

where the valons are constituent or 'dressed' valence quarks surrounded by a sea of gluons and $q\bar{q}$ pairs. For example, a nucleon is made of just three valons, the detailed internal structure of which cannot be resolved at low Q^2. In deep inelastic lepton scattering the virtual photon can resolve the partons in a valon. Their structure functions are assumed to be universal with an evolution in Q^2 given by lowest-order perturbative QCD with a δ function input at low Q^2. If this can be accepted, deep inelastic lepton scattering and lepton pair production data may be used to determine the valon structure function of an initial proton or pion, which in turn predicts the recombination of valons into the final hadron. The calculations have been carried out for pp $\rightarrow \pi^+$X (Hwa and Zahir 1981).

To summarise, the theoretical interpretation of low p_T fragmentation is still open to debate, and continues to attract a lot of controversy. We have seen that the existing parton models divide essentially into the quark fragmentation and the quark recombination approaches. Put simply, in fragmentation models a valence quark is 'held back' and the inclusive distribution is described by the fragmentation functions, $D_q^h(x)$, of the remaining fast forward-moving quark (or diquark) system. But in the recombination approach the inclusive distribution is given by the structure function, $f^q(x)$, of a fast forward-moving valence quark which recombines with a slow sea quark to produce the outgoing hadron. Although these seem to be contradictory viewpoints, there have been claims that from the standpoint of the dual topological unitarisation scheme they are essentially equivalent parton model interpretations of the same dual cylinder component (Cohen-Tannoudji *et al* 1980). In any case all these models are at best only rather crude first approximations to the underlying QCD mechanisms. We shall have more to say about this in chapters 10 and 11.

6.2   Counting rules for low p_T fragmentation

The inclusive distributions at high energy and low p_T are frequently fitted to the form $d\sigma/dx = A(1-x)^\beta$. Some of the observed values of β are listed in figure 56, which is taken from Bobbink *et al* (1980). Further results can be found in Denegri *et al* (1981).

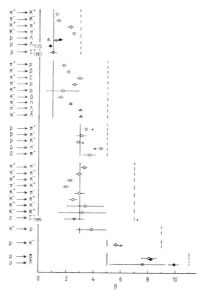

Figure 56. The exponents β obtained from fitting data for various reactions to the form $d\sigma/dx \sim (1-x)^\beta$. The lines are the counting rule predictions described in the text. The figure is from Bobbink *et al* (1980).

The quark counting rules (see, for example, Blankenbecler and Brodsky 1974) predict that as $x \to 1$ in the process $AB \to CX$

$$\frac{d\sigma^{(A \to C)}}{dx} \sim (1-x)^{2n_s - 1} \tag{6.7}$$

where n_s is the minimum number of spectator quarks involved in the transition $A \to C$. An example was given in (6.3). The more spectators share the initial momentum the smaller is the chance of producing a hadron with a large fraction of the momentum.

The formulation of the counting rule has evolved in the light of experiment. The number of relevant spectators depends on whether the hadronic interaction $AB \to CX$ proceeds dominantly via quark or gluon exchange; on whether we count sea, as well as valence, quarks among the spectators; and, if so, how this is to be done. For example, consider $pp \to \pi^+X$; the lowest Fock state of the proton which fragments into a π^+ is $|uudd\bar{d}\rangle$. Figure 57 shows the fragmentation proceeding accompanied by either gluon exchange or quark exchange. In the former case there are three spectators yielding a $(1-x)^5$ behaviour, and in the latter two spectators giving a $(1-x)^3$ behaviour which is in better agreement with the data.

Figure 57. $p \to \pi^+$ fragmentation via gluon or quark exchange diagrams.

However, quark exchange leads to a dramatic long-range correlation between the fragmentation behaviour of the beam and target in a double inclusive reaction such as $pp \to \pi^+ \pi^+ X$. For example, for d quark exchange in figure 57 the lowest Fock state of the other incident proton is $|uudd\bar{d}d\bar{d}\rangle$ which fragments into a π^+ with four spectators. So with quark exchange the double inclusive cross section is

$$\frac{d^2\sigma(pp \to \pi_1^+ \pi_2^+ X)}{dx_1 \, dx_2} \sim [(1-x_1)^3 (1-x_2)^7 + 1 \leftrightarrow 2].$$

However, the recent ISR data for $pp \to \pi^+ \pi^+ X$ show a complete absence of such a correlation (Bobbink *et al* 1980).

Indeed, this lack of correlation is an essential feature of the models we discussed earlier for low p_T fragmentation. It is assumed that one incident hadron fragments independently of the other. An important test of this is that the ratio of $\pi^\pm$ inclusive production, $hp \to \pi^+ X / hp \to \pi^- X$ in the proton fragmentation region should be independent of whether the beam hadron, h, is a p, π^+, π^-, K^+, etc. Experimentally this is found to be the case above about 200 GeV/c, but at lower momenta a component that vanishes as $s^{-1/2}$ is required (Kittel 1981), which may be due to quark exchange contributions.

Returning to the counting rule we see that for gluon exchange we expect $\beta = 5$ for $p \to \pi^+$ fragmentation. The predictions for other fragmentation processes are shown by the broken lines in figure 56 and clearly disagree with the observed exponents. Prompted by this discrepancy, a physically reasonable suggestion is to count only valence quarks as spectators. This is motivated by the recombination model in which the role of sea quarks in sharing the momentum among spectators is reduced, relative to that of the valence quarks, by the peaking of the sea structure functions at small x. The full lines on figure 56 follow from this valence-quark-spectator counting rule, and are seen to be in better agreement with experiment.

Recently Gunion (1979) has reconsidered the counting rules from the viewpoint of lowest-order QCD. The reason why perturbative QCD may be applicable to low p_T fragmentation is illustrated by the fragmentation of a π into a valence quark (see figure 58). Initially the q and $\bar{q}$ have, on the average, equal fractions of the pion's momentum. In the lowest order, the way to obtain a quark with momentum fraction x near 1 is to transfer momentum from the $\bar{q}$ to the q via gluon exchange. With the

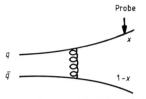

Figure 58. A π meson fragmenting into a valence quark, with one hard gluon leaving a single spectator.

pion and $\bar{q}$ on mass shell, the four-momentum squared of the quark q is

$$k^2 = -\frac{k_T^2 + m^2(x)}{1-x}$$

where m depends on the particle masses and is regular as $x \to 1$. Thus as $x \to 1$ the probed quark is far off mass shell and the momentum measurement acts as a short-distance probe. Neglecting spin, Gunion shows this contribution to the quark structure function of the pion is

$$f_\pi^v(x) \sim (1-x)$$

as $x \to 1$. Brodsky and Lepage (1979a,b) have considered the leading log QCD diagrams in this limit and shown that this lowest-order power law is not modified, although there are logarithmic modifications.

For the sea quark distribution of the pion, Gunion uses the lowest-order diagram of figure 59(a) which yields

$$f_\pi^s(x) \sim (1-x)^3$$

since two hard gluons are needed to transfer all the momentum to the sea quark.

(a) $\hspace{8cm}$ (b)

Figure 59. Possible diagrams for a π meson to fragment into a 'sea' quark, leaving three spectators.

This is to be compared with the hadronised four-quark diagram (figure 59(b)), which contains the q_s initially, which needs three hard gluons, and gives $(1-x)^5$ since there are three spectators. Applying this approach to $pp \to \pi^+ X$, the lowest-order diagram is figure 60 which yields

$$\frac{d\sigma^{(p \to \pi^+)}}{dx} \sim (1-x)^3.$$

Figure 60. Pair creation diagram for $p \to \pi^+$ fragmentation, with two hard gluons.

The general rule, originally formulated for QED by Blankenbecler *et al* (1975), is that the $x \to 1$ behaviour of a diagram is

$$(1-x)^{2n_H+n_{PL}-1} \tag{6.8}$$

(where n_H is the number of spectators originating from the hadronic wavefunction and n_{PL} is the number of spectators associated with the point-like pair creation process) which corresponds to the count of hard gluons given above.

This rule agrees rather well with the data (see Gunion 1979). In particular exotic fragmentation processes are relatively weakly suppressed compared to the original spectator counting rule. For example, for $\pi^+ \to K^-$ fragmentation of figure 61 it gives $(1-x)^3$ as compared to $(1-x)^7$. In fact, the predictions of Gunion's rule are very similar to the suggestion of counting just the valence quark spectators of the hadronised state. For $\pi^+ \to K^-$ this also gives $\beta = 3$ (see figure 56).

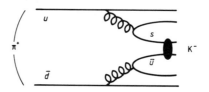

Figure 61. Pair creation diagram for $\pi^+ \to K^-$ fragmentation ($n_{PL} = 4$, $n_H = 0$).

Despite the empirical support for the use of these point-like pair creation QCD diagrams, with minimal suppression as $x \to 1$, their theoretical justification is far from clear. There appears to be no good reason why the created quarks should not interact with the other quarks so that the bound state momentum is redistributed, giving back the rule based on counting the spectators in the initial hadronised state.

Finally we note that the application of parton ideas to low p_T fragmentation should be restricted to $x \lesssim 0.8$. For larger x values triple-Regge phenomenology will dominate, as we shall discuss in chapter 8.

7

EXCHANGE FORCES

7.1 Regge poles

In § 1.5 we remarked that the strong interaction between composite hadrons is a mutual colour polarisation effect, somewhat analogous to the interatomic forces which produce molecules. It involves the exchange of quarks and gluons between the hadrons, but to preserve the colour neutrality (whiteness) of the particles no net colour can be exchanged, only colourless clusters of partons. Hence the longest-range component of the force is provided by exchanging the lightest available hadron with the appropriate flavour quantum numbers, as in figure 8(a).

For the process AB → CD (figure 62) s, defined in (2.1), is the square of the centre-of-mass energy while t in (2.10) gives the scattering angle. However, the

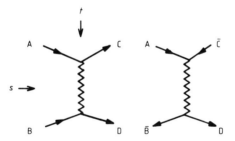

Figure 62. The s-channel (AB → CD) and t-channel (A$\bar{\text{C}}$ → $\bar{\text{B}}$D) processes which are related by crossing. The particles which are interchanged (B, C) become their antiparticles in order to conserve quantum numbers. The particles exchanged in the process AB → CD must have the quantum numbers of A$\bar{\text{C}}$ and $\bar{\text{B}}$D.

exchanged particle is formed in the t-channel process A$\bar{\text{C}}$ → $\bar{\text{B}}$D in which t gives the energy and s the scattering angle. Thus for the simplest case where $m_A = m_B = m_C = m_D \equiv m$, (2.10) with s and t interchanged gives

$$\cos \theta_t = 1 + \frac{2s}{t - 4m^2} \qquad (7.1)$$

for the t channel. These s and t channels are related by crossing and share a common scattering amplitude, but involve different regions of the s and t variables (see, for example, Collins 1977).

The old-fashioned one-pion-exchange (OPE) approximation to the nuclear force consisted of representing the scattering amplitude by the t-channel pion propagator pole

$$A(s, t) \sim \frac{1}{m_\pi^2 - t}. \qquad (7.2)$$

There are two related difficulties with this approach, however. First the pion is just the lightest of many particles with the required flavour quantum numbers, and to obtain the full nuclear interaction one must include the others, such as ρ, ω, f and A_2 as well as multiparticle exchange (2π, 3π, etc) which provide shorter-range contributions. Secondly, and more fundamentally, (7.2) is really just an approximation to the S wave ($l = 0$ partial wave) in the t channel (because it is independent of s and hence of $\cos\theta_t$, from (7.1)). This partial-wave decomposition is fine for the t-channel process $A\bar{C} \rightarrow \bar{B}D$, but it will not do in the s-channel $AB \rightarrow CD$ where we want to use it. This is because any singularity in s gives rise to a divergence of the t-channel partial-wave series. For example, an s-channel particle pole

$$\frac{1}{m^2 - s} = \frac{1}{m^2}\left[1 + \frac{s}{m^2} + \left(\frac{s}{m^2}\right)^2 + \ldots\right] \tag{7.3}$$

diverges as $s \rightarrow m^2$. Hence the t-channel partial-wave series

$$A(s, t) = \sum_{l=0}^{\infty} (2l + 1)A_l(t)P_l(\cos\theta_t) \tag{7.4}$$

(where $A_l(t)$ is the partial-wave amplitude and P_l is a Legendre polynomial) will also diverge as $s \rightarrow m^2$ because of (7.1). There will certainly be many singularities in s (bound state and resonance poles, threshold branch points, etc) intervening between the t- and s-channel physical regions, and so the approximation (7.2) will break down before the s-channel region is reached. Instead we need a method of summing the full partial-wave series. This will not only enable us to overcome the divergence problem, but will also allow the simultaneous inclusion of all particles which lie on a given Regge trajectory like figure 9, and which occur in different t-channel partial waves.

A general way of achieving this was suggested by Sommerfeld (1949) (following Watson 1918) and is developed in detail in, for example, Collins (1977). For our purpose it will suffice to consider the particles lying on a single linear trajectory

$$\alpha(t) = \alpha_0 + \alpha't \tag{7.5}$$

such that $\alpha(t)$ passes through integer values of l at $t = m_l^2$ ($l = 0, 1, 2, \ldots$). The pole in the lth partial wave then takes the form

$$A_l(t) \simeq \frac{\beta(t)}{l - \alpha(t)} \simeq \frac{\beta(t)}{\alpha'(m_l^2 - t)} \tag{7.6}$$

i.e. there is a 'Regge pole' in the partial-wave amplitude at $l = \alpha(t)$, and $\beta(t)$ is the residue function specifying the coupling of the pole to the external particles. The contribution of the trajectory to the amplitude is then, from (7.6) substituted in (7.4),

$$A(s, t) \simeq \sum_{l=0}^{\infty} (2l + 1)\frac{\beta(t)}{l - \alpha(t)}P_l(\cos\theta_t). \tag{7.7}$$

Since the asymptotic form of $P_l(\cos\theta) \rightarrow (\cos\theta)^l$ as $\cos\theta \rightarrow \infty$, we find that (Perl 1974, p395)

$$A(s, t) \underset{s \rightarrow \infty}{\sim} \sum_{l=0}^{\infty} \frac{\beta(t)(\cos\theta_t)^l}{l - \alpha(t)} \sim \beta(t)(\cos\theta_t)^{\alpha(t)} \sim \beta(t)s^{\alpha(t)} \tag{7.8}$$

at fixed t, from (7.1).

Equation (7.8) is the characteristic Regge-pole asymptotic power behaviour of the scattering amplitude as a function of s at fixed t stemming from the exchange of a Regge trajectory of composite particles (Regge 1959, 1960). It predicts that in a two-body process

$$\frac{d\sigma}{dt} \sim \frac{1}{s^2}|A(s,t)|^2 \sim F(t)\left(\frac{s}{s_0}\right)^{2\alpha(t)-2} \tag{7.9}$$

where $\alpha(t)$ is the leading Regge trajectory which can be exchanged, and $s_0 \approx 1\ \text{GeV}^2$ is the hadron mass scale. This has been well verified in many processes (Collins 1977, Irving and Worden 1977).

The trajectory $\alpha(t)$ is readily determined by plotting $\log(d\sigma/dt)$ as a function of $\log s$ at each t value. For example, in $\pi^- p \to \pi^0 n$ the t-channel process $\pi^-\pi^0 \to \bar{p}n$ has the flavour quantum numbers $I = 1$, $P = G = +$, i.e. of the ρ meson in table 2, so the ρ trajectory of figure 9 can be exchanged. But in the s channel $-t$ is the square of the momentum transferred and so (7.8) gives us the continuation of this trajectory to negative values of t, as shown in figure 63.

If (7.5) is substituted into (7.9) we find

$$\frac{d\sigma}{dt} \sim F(t)\left(\frac{s}{s_0}\right)^{2\alpha_0-2} \exp\left[2\alpha'\log(s/s_0)t\right] \tag{7.10}$$

and so the forward peak in $|t|$ becomes sharper ('shrinks') as $\log s$ increases. This

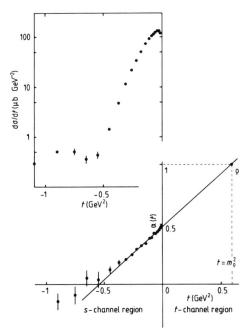

Figure 63. The $\pi^- p \to \pi^0 n$ differential cross section for beam momentum 20.8 GeV/c (Barnes *et al* 1976). The lower plot compares the values of $\alpha(t)$ (with $t \leq 0$) obtained by fitting to $\pi^- p \to \pi^0 n$ data in the momentum range 20–200 GeV/c with the extrapolation of the linear ρ trajectory of figure 9.

shrinkage, which is due to the slope of the Regge trajectory, is a characteristic feature of Regge poles.

However, (7.8) only indicates the asymptotic s behaviour of a Regge pole. A more complete expression, which exhibits the principal features of Reggeon exchange, such as figure 62, is (Collins 1977)

$$A(s, t) = \gamma_{AC}(t)\gamma_{BD}(t)\frac{\exp[-i\pi\alpha(t)]+\mathscr{S}}{\sin \pi\alpha(t)}\frac{1}{\Gamma(\alpha(t))}\left(\frac{s}{s_0}\right)^{\alpha(t)}. \qquad (7.11)$$

Here $\gamma_{AC}(t)$ represents the coupling of the trajectory to particles A and C at the upper vertex, and γ_{BD} is the lower vertex. This factorisation property of the coupling is well verified. The factor $[\sin \pi\alpha(t)]^{-1}$ is the Reggeon propagator and produces the required resonance poles in t whenever $\alpha(t)$ passes through an integer, as in (7.6). $\mathscr{S}$ is called the 'signature' and takes the values $\mathscr{S} = \pm1$ for even/odd signature trajectories. It arises because the Reggeon is really the sum of two terms, as in figure 64, and we get

$$(-s)^{\alpha(t)}+\mathscr{S}(-u)^{\alpha(t)} \to s^{\alpha(t)}\{\exp[-i\pi\alpha(t)]+\mathscr{S}\} \qquad (7.12)$$

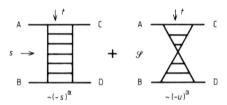

Figure 64. A Reggeon (of signature $\mathscr{S}$) is the sum, (7.12), of planar s–t and planar u–t contributions, respectively. Planar diagrams are ones in which the particle lines do not cross; the second diagram becomes planar in u, t if it is untwisted by interchanging B and D (see (2.1) and (2.11)). The rungs of the ladders represent the sum over all physical intermediate states in the $s(u)$ channel and, through unitarity, build up the imaginary part of the amplitude in $s(u)$.

as $s \to \infty$ at fixed t from (2.12). It means that (7.11) only has poles for even/odd integer values of $\alpha(t)$ respectively, and so even and odd partial waves (with their different symmetry properties under $\cos \theta_t \leftrightarrow -\cos \theta_t$, i.e. $s \leftrightarrow u$) have separate trajectories. This signature factor has the important consequence that, since $\gamma(t)$, $\alpha(t)$ are expected to be real functions of t for $t < 0$, the phase of (7.11) is given by

$$\rho \equiv \frac{\mathrm{Re}\,\{A\}}{\mathrm{Im}\,\{A\}} = -\frac{\cos \pi\alpha + \mathscr{S}}{\sin \pi\alpha}. \qquad (7.13)$$

This phase is required for consistency with fixed-t dispersion relations, and is well verified experimentally. The Γ function is included in (7.11) to ensure that there are no unphysical resonance poles at negative values of $\alpha(t)$. Since

$$\frac{1}{\Gamma(\alpha)} = -\frac{\sin \pi\alpha}{\pi}\Gamma(1-\alpha) \qquad (7.14)$$

it is evident that this function vanishes for $\alpha = 0, -1, -2, \ldots$, and so cancels the 'nonsense' poles of the Reggeon propagator. The ρ meson has spin 1 and so its trajectory has $\mathscr{S} = -1$. Hence when $\alpha(t) \to 0$, which we see from figure 63 happens for $t \approx -0.5 \,\mathrm{GeV}^2$, (7.11) vanishes. The dip in the $\pi^-p \to \pi^0 n$ differential cross section is usually regarded as a verification of the nonsense decoupling at this point. (There are, however, alternative explanations involving cuts, see chapter 9.)

Despite the fact that the ρ and A_2 trajectories have opposite signature ($\mathscr{S} = -1$ and $+1$, respectively, giving leading particles with spin 1 and 2) these two trajectories look like a single trajectory in figure 9, with particles at every positive integer value of s. The observed trajectories are thus exchange-degenerate. The reason for this is that the QCD potential between the quark and antiquark is the same for the ρ and A_2 particles, which differ only through the orbital angular momentum of the $q\bar{q}$ system—see § 1.5. If their couplings are equal too then in processes like $K^- p \rightarrow \bar{K}^0 n$ and $K^+ n \rightarrow K^0 p$ where we have $A_2 \pm \rho$ exchange (the sign change occurring because ρ has negative C), the respective phases will be given by

$$(\exp[-i\pi\alpha(t)] + 1 \pm \{\exp[-i\pi\alpha(t)] - 1\}) = 2 \exp[-i\pi\alpha(t)], 1 \qquad (7.15)$$

so the latter reaction should have a real phase, which is certainly experimentally confirmed (via the optical theorem), while the former has a phase which changes with $\alpha(t)$. But their differential cross sections, which depend on $|A|^2$ only, should be identical. The data are in reasonable agreement with these predictions (see, for example, Irving and Worden 1977).

The optical theorem illustrated in figure 65 (see Collins 1977) gives

$$\sigma_T(AB) \sim \frac{1}{s} \operatorname{Im} A(AB \rightarrow AB, s, t = 0) \sim s^{\alpha(0)-1} \qquad (7.16)$$

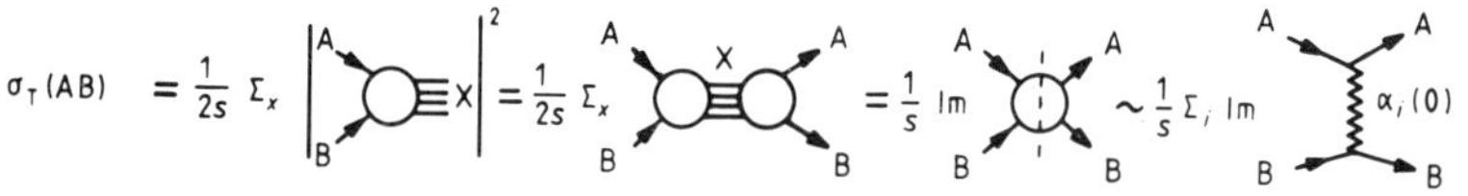

Figure 65. Pictorial representation of the optical theorem, which relates the AB total cross section to the imaginary part of the forward amplitude for elastic AB scattering. The third equality follows from unitarity.

where α is the leading trajectory which can be exchanged in elastic scattering. In figure 10 we see that at high energy all the total cross sections are nearly constant with energy, which implies that $\alpha(0) \approx 1$ in (7.16). This is not true of the meson trajectories in figure 9, which have $\alpha(0) \approx \frac{1}{2}$, nor of any others presently known. The leading trajectory exchanged in elastic scattering has the quantum numbers of the t-channel $A\bar{A} \rightarrow B\bar{B}$, which is flavourless and should be universal. It seems reasonable to suppose therefore that, unlike ordinary meson trajectories which contain valence quarks, the flavourless exchange in elastic scattering is made of gluons. Since no net colour can be transferred the minimum contribution is the two-gluon exchange box of figure 23. As we noted in chapter 2 this trajectory is called the Pomeron P. Its properties will be discussed in chapter 9.

In addition to the exchange of a single Reggeon, as in figure 62, it is also possible to exchange two or more Reggeons simultaneously. This gives rise to Regge cuts, which become important at large values of $-t$, as we shall see in chapter 9. For small $-t$ however, the Regge poles are dominant and are an indispensable tool for the analysis of hadron cross sections. By incorporating SU(3) flavour symmetry into the Regge residues, and assuming that the s dependence is given by extrapolations of Regge trajectories which are linear functions of t like (7.5) passing through the known resonances (like figure 63), and that the amplitude phase is given by the signature factor as in (7.13), one can predict quite well the behaviour of hadron scattering processes at low $|t|$ and high s. The scope of this success is summarised in, for example, Collins (1977) and Irving and Worden (1977).

We next want to look at the parton description of Regge exchanges.

7.2 Duality and quark line diagrams

In the parton model flavour is carried by the quarks and hence the exchange of flavour (e.g. the exchange of charge in $\pi^- p \to \pi^0 n$ discussed above) can be represented by quark line diagrams showing just the valence quarks of the hadrons which are involved. Some examples are shown in figures 66(a) and (d).

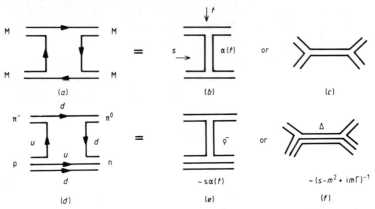

Figure 66. Duality diagrams for MM → MM and $\pi^- p \to \pi^0 n$, representing both Regge exchanges (b) and (e) and resonances (c) and (f).

These figures also illustrate the fact that the same diagram can be used to describe (i) the exchange of a Reggeon in the t channel, giving an amplitude behaving like $s^{\alpha(t)}$ as $s \to \infty$, and (ii) s-channel resonances which give Breit–Wigner poles in the amplitude

$$A \underset{s \to m^2}{\sim} \frac{m\Gamma}{s - m^2 + im\Gamma} \tag{7.17}$$

where m is the resonance mass and Γ is its width. The diagram represents not just a single resonance, however, but all the (infinite number of?) resonances which can be formed with the same valence quarks, including both orbital and radial excitations.

The hypothesis of duality (see, for example, Collins (1977) for a review and references) is that these two types of diagrams ((b) and (c), or (e) and (f)) are in fact equivalent (dual to each other) and really represent just a single contribution to the scattering amplitude. This means that the sum of all the many Regge exchanges represented by figure 66(b) is identical to the sum of all the resonances represented by figure 66(c). For $\alpha(t) > -1$ this can only be possible if the number of resonances is divergent, since individually each resonance $\sim s^{-1}$ as in (7.17). A concrete mathematical realisation of this hypothesis is given by the Veneziano model (1968) for the scattering amplitude, viz

$$V(s, t) = \frac{\gamma^2}{\pi} \frac{\Gamma(1 - \alpha(s))\Gamma(1 - \alpha(t))}{\Gamma(1 - \alpha(s) - \alpha(t))}. \tag{7.18}$$

From the inverse of (7.14) it is evident that $\Gamma(1 - \alpha(t))$ gives rise to resonance poles in s whenever $\alpha(s)$ passes through a positive integer, and similarly there are poles in t whenever $\alpha(t)$ is a positive integer, but the Γ function in the denominator ensures that there are no double poles when $\alpha(s)$ and $\alpha(t)$ are integers simultaneously.

From Stirling's formula (Magnus and Oberhettinger 1949 p4) we find that

$$\frac{\Gamma(x+a)}{\Gamma(x+b)} \xrightarrow[x\to\infty]{} x^{a-b} \tag{7.19}$$

(except in a wedge along the negative x axis where poles appear at integer x). So if we have a linear trajectory like (7.5)

$$V(s,t) \xrightarrow[\substack{s\to\infty \\ t\,\text{fixed}}]{} \frac{\gamma^2(-\alpha's)^{\alpha(t)}}{\Gamma[\alpha(t)]\sin\pi\alpha(t)} = \frac{\gamma^2(\alpha's)^{\alpha(t)}}{\Gamma[\alpha(t)]\sin\pi\alpha(t)} \exp[-i\pi\alpha(t)] \tag{7.20}$$

which may be compared with (7.11).

The model (7.18) thus has the following properties. (a) It is manifestly crossing symmetric with resonance poles and Regge asymptotic behaviour in both s and t. (b) It has the nonsense Γ factor to remove the unphysical poles at negative integer values of α. (c) The scale factor s_0 in (7.11) has been replaced by $(\alpha')^{-1}$ in (7.20), and we note from figure 9 that $(\alpha')^{-1} \approx 1\,\text{GeV}^2$, the hadron mass scale. (d) The asymptotic phase is $\exp[-i\pi\alpha(t)]$, which gives a positive imaginary part for $s>0$. To reproduce the phase of (7.11) we need the sum of two terms, like figure 64, viz

$$[V(s,t)+\mathcal{S}V(u,t)] \xrightarrow[\substack{s\to\infty \\ t\,\text{fixed}}]{} \frac{\gamma^2(\alpha's)^{\alpha(t)}}{\Gamma[\alpha(t)]\sin\pi\alpha(t)}\{\exp[-i\pi\alpha(t)]+\mathcal{S}\}. \tag{7.21}$$

(e) In an exotic process such as $\pi^+\pi^+ \to \pi^+\pi^+$, as there are no charge-two meson resonances there are no poles in s and hence only the second term of (7.21) appears and the phase is real. This is because the imaginary parts of the ρ and f exchanges cancel by exchange degeneracy, as in (7.15). The imaginary part of the amplitude, and hence from (7.16) $\sigma_T(\pi^+\pi^+)$, is given entirely by the gluon-exchange Pomeron (P). On the other hand, in $\pi^+\pi^- \to \pi^+\pi^-$, which does have s-channel resonances, the ρ and f contributions add (+ sign in (7.15)) and σ_T involves these trajectories as well as P. Similarly $pp \to pp$ is exotic but $\bar{p}p \to \bar{p}p$ is not and so the total cross sections are given by

$$\sigma_T(pp) = P+f-\omega+A_2-\rho \sim As^{\alpha_P(0)-1} \tag{7.22}$$

$$\sigma_T(\bar{p}p) = P+f+\omega+A_2+\rho \sim As^{\alpha_P(0)-1} + Bs^{\alpha_R(0)-1}$$

the sign changes stemming from the fact that ω and ρ are odd under charge conjugation ($p \leftrightarrow \bar{p}$). With exchange degeneracy $f=\omega$ and $A_2=\rho$, and so only P contributes to pp, whereas the four (degenerate) trajectories of figure 9, with $\alpha_R(0) \approx \frac{1}{2}$, do not cancel in $\bar{p}p$. This explains why $\sigma_T(\bar{p}p)$ is higher than $\sigma_T(pp)$ at low energies in figure 10, but falls to meet it like $s^{-1/2}$ at high s.

As reviewed in, for example, Jacob (1974) and Collins (1977), this duality idea, though it lacks any very fundamental justification at the present time, enables us to summarise conveniently many of the properties of Regge exchanges, including their flavour dependence, exchange degeneracy, etc. Because of this it is useful to formalise the rules for drawing duality quark line diagrams. These are generally called Zweig's rules (Rosner 1969) and are:

(i) The flavour of each particle is represented by its valence quarks, with a positive arrow for a quark and negative for a $\bar{q}$, so a meson is ⇉ and a baryon is ⇛ .

(ii) Each hadronic process is represented by a connected planar diagram which can be cut at any intermediate stage by lines representing just a single meson or baryon. This ensures flavour conservation, no exotic intermediate states and exchange degeneracy.

(iii) No quark line begins and ends on the same particle (this would be a sea quark).

The diagrams of figure 66 accord with these rules. An interesting example is provided by $\pi^- p \to Vn$ where V is a vector meson; ω, φ, ψ or Υ. We see in figure 67

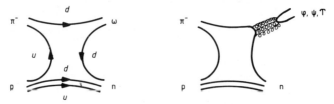

Figure 67. Allowed and 'forbidden' duality diagrams for $\pi^- p \to Vn$.

that only the first has an allowed diagram, while the others require the creation of a sea $q\bar{q}$ pair, and these processes are highly suppressed. Similarly the decay $\psi^*(4415) \to D\bar{D}$ is allowed (figure 68), but $\psi(3100)$ is below the $D\bar{D}$ threshold so its hadronic decays violate Zweig's rules and are highly suppressed. This is why the latter is very narrow (long-lived) while the former is not. In the $\psi^*(4415)$ the excited $c\bar{c}$ pair can easily lose energy by creating a $u\bar{u}$ or $d\bar{d}$ pair in the vacuum to produce the lower-mass D and $\bar{D}$ mesons. But in the $c\bar{c}$ ground state, $\psi(3100)$, they do not have enough energy for this, and instead must first annihilate into at least three virtual gluons (to conserve colour and charge conjugation) which create the final-state hadrons when they attempt to escape confinement. The probability for this is proportional to $[\alpha_s(m_\psi^2)]^3$, and since α_s is quite small ($\simeq 0.3$) the decay is highly suppressed.

Figure 68. Allowed and 'forbidden' duality diagrams for $\psi(c\bar{c})$ decays.

The representation of hadronic processes by these quark line diagrams is a useful way of summarising all the main phenomenological features such as resonance excitation and decay, Regge exchanges and the phases of amplitudes. They do not, however, include any threshold branch points, or Regge cuts. One consequence of this is that the poles in s in (7.18) lie on the real s axis, because $\alpha(s)$ is a real function, and so the amplitude becomes infinite whenever $\alpha(s)$ is an integer (at $s = m^2$ say). This is of course completely unphysical since the magnitude of the partial wave amplitude is bounded by 1 by unitarity (cf (9.1) below). The resonance poles should be displaced off the real s axis as in (7.17) giving a large but finite amplitude at $s = m^2$. The Veneziano model is thus not unitary, and attempts to construct realistic dual models consistent with unitarity have not been very successful. Despite this the Veneziano model is a remarkably good first approximation to the scattering amplitude, incorporating all the resonance and exchange poles on Regge trajectories. It is thus essential to try and gain some understanding of how Reggeons may be generated dynamically in QCD.

7.3 The dynamics of Regge trajectories

The particles which lie on a meson Regge trajectory like figure 9 are $q\bar{q}$ bound states. To calculate a Regge trajectory from first principles would thus require a solution to the QCD confinement problem, which is far beyond our theoretical competence at the moment. We can, however, attempt an educated guess as to how the dynamics may work, based on experience with much simpler models.

In § 1.3 we remarked how non-relativistic potential models, with confining potentials which are increasing functions of the separation r at large r, can account for the charmonium (ψ) and beauty (Y) states. The radial Schrödinger equation contains the effective potential (1.14), in which the first term is the attractive interaction potential and the second term is the centrifugal repulsion representing the difficulty of holding high-l states together. Usually this repulsion comes to dominate at large l and so high-l bound states cannot be formed. In states of large l the constituents spend much of their time at large r, where the potential is weak. However, confining potentials are an exception to this because $V(r)$ increases with r and so the bound state can still be held together. High-l states are, of course, heavier because the constituents have greater internal kinetic energy.

A simple example is the three-dimensional harmonic oscillator potential, $V(r) = \lambda r^2$, whose energy eigenvalues are (Morse and Feshbach 1953 p1662)

$$E_{nl} = (2n_r + l + \tfrac{3}{2})\hbar\omega_c \tag{7.23}$$

where $n_r = 0, 1, 2, \ldots$, is the radial quantum number and $2\pi\omega_c = (\lambda/m)^{1/2}$, m being the quark mass. We thus find a linear trajectory for l against E. More generally it is found that for $V \sim r^n$ the asymptotic form of the trajectory is (Quigg and Rosner 1979)

$$l \sim E^{(n+2)/2n}. \tag{7.24}$$

Note that with $n = -1$ this gives $l \sim E^{-1/2}$, in agreement with the Rydberg formula for the hydrogen atom $E_{nl} \sim (n_r + l + 1)^{-2}$. The observed linearity of l with M^2 in figure 9 might suggest that $n = \tfrac{2}{3}$. However, as we discussed in chapter 1, it is only possible to take these potential models seriously when the system is non-relativistic ($E \ll mc^2$) so these high-E limits are probably of limited relevance. The important point is that the increase of l with M^2 is evidence for the increase of the interaction strength with r, and hence for at least partial confinement.

Another, somewhat more realistic, model which we can use for Regge trajectories is based on Feynman perturbation field theory (Eden et al 1966, Collins 1977, Polkinghorne 1980). The results for the interactions of scalar particles in φ^3 theory are very well known. The basic interaction amplitude (figure 69(a)) is single scalar exchange $(s - m^2)^{-1}$, where m is the mass, and so as $s \to \infty$ it behaves like s^{-1}. If we include further exchanges it is found that the leading $\log s$ form of the amplitude is

(a)

Figure 69. Ladder diagrams for scalar particles which sum to give $s^{\alpha(t)}$ behaviour.

given by the ladder diagrams, and that an n-rung ladder has the asymptotic form

$$\sim \frac{[K(t)\log s]^{n-1}}{s(n-1)!} \tag{7.25}$$

where $K(t)$ arises from the box diagram loop integration. Summing the diagrams with all possible numbers of rungs gives

$$A(s,t) \sim \sum_{n=1}^{\infty} \frac{[K(t)\log s]^{n-1}}{s(n-1)!} \sim s^{\alpha(t)} \tag{7.26}$$

where $\alpha(t) \equiv -1 + K(t)$. It can be verified in perturbation theory that more complicated diagrams (including those in which the rungs cross over each other) either do not contribute to the leading $\log s$ behaviour, or constitute renormalisation effects. The leading behaviour always comes from ladder diagrams like figure 9 in which the rungs are coupled in the order of their rapidities (De Tar 1971). So if one is willing to assume that this set of diagrams is the correct one to obtain the leading behaviour, we have generated a Regge trajectory by summing ladders. In this model $K(t) > 0$ and $\to 0$ as $t \to -\infty$, so that $\alpha(t) \xrightarrow[-t \to \infty]{} -1$. This is because the Born approximation (figure 69(a)), which controls the large t limit, behaves like s^{-1}.

There are several problems with applying these ideas to QCD with gluon exchanges between quarks. First, we have a spin-1 gluon propagator with s dependence in the numerator. Secondly, the quarks have spin $\frac{1}{2}$ so the total angular momentum of the Born diagram is the vector sum of their spins and the orbital angular momentum. Thirdly, we are dealing with massless gluons in an IR-divergent perturbation theory. And finally, and perhaps most serious, we have a non-Abelian UV-divergent confining theory, so that the quarks and gluons in the intermediate states do not represent the unitarity structure properly. The intermediate states should really be physical particles. Some subtle cancellations are known to occur (Grisaru *et al* 1973, Bartels 1980) but the complete solution is not known.

We shall simply ignore most of these problems, and begin by observing that the basic Born diagram (figure 70(a)) has the scaling form $\sim t/s$ for s, $-t \to \infty$, s being the gluon propagator denominator (see table 3—the roles of s and t are interchanged). If we then assume that the multi-gluon exchanges of figure 70 Reggeise like those

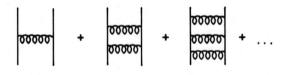

(a)

Figure 70. Quark–gluon ladder diagrams which are speculated to sum to give a $ts^{\alpha(t)}$ behaviour.

of figure 69 we may expect the behaviour $\sim ts^{\alpha(t)}$ with $\alpha(-\infty) = -1$, though, because of the cancellations noted above, it is certainly not obvious that the trajectory must end up at -1. Indeed, figure 63 indicates that it may continue straight down (see Collins *et al* 1968). The suggestion that the trajectory should eventually asymptote to a negative integer (Blankenbecler *et al* 1973, 1974) stems from the fact that hard scattering (2.24) controls the large-angle limit, as discussed in § 2.3, and it is hoped that this matches smoothly onto the Regge limit. (There is clearly a strong similarity between figure 66 and the constituent interchange model of figure 38.) If so, then

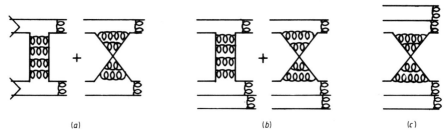

Figure 71. Quark line diagrams (with gluon insertions) which lead to the asymptotic behaviour for (a) MM, (b) MB and (c) BB hard scattering.

for meson scattering, MM → MM, the amplitude may take the form (figure 71(a))

$$A(s, t) \underset{s \to \infty}{\sim} F_M^2(t) t s^{\alpha(t)} \underset{|t| \gg m^2}{\sim} \frac{1}{t} s^{\alpha(t)} \tag{7.27}$$

where, as described in § 2.3, the form factor $F_M(t)$ gives the probability of the meson re-forming after the hard scattering, and its asymptotic behaviour is given by (2.22). Then

$$\frac{d\sigma}{dt} \underset{s \to \infty}{\sim} \frac{1}{s^2} |A|^2 \sim \frac{1}{t^2} s^{2\alpha(t)-2}. \tag{7.28}$$

But in the fixed-angle limit, s, $-t \to \infty$, s/t fixed, we have $\alpha(t) \to -1$ and so $d\sigma/dt \sim s^{-6} f(t/s)$ in agreement with (2.26). Similarly for MB and BB scattering we find

$$A(MB \to MB) \underset{s \to \infty}{\sim} F_M(t) F_B(t) t s^{\alpha(t)} \qquad \text{so} \qquad \frac{d\sigma}{dt} \underset{s,-t \to \infty}{\sim} s^{-8} f\left(\frac{t}{s}\right)$$

$$\tag{7.29}$$

$$A(BB \to BB) \underset{s \to \infty}{\sim} F_B^2(t) t s^{\alpha(t)} \qquad \text{so} \qquad \frac{d\sigma}{dt} \underset{s,-t \to \infty}{\sim} s^{-10} f\left(\frac{t}{s}\right)$$

from (2.22), in accord with (2.27) and (2.28). Note that in BB scattering only the t–u planar diagram (figure 71(c)) is possible, giving the real phase of (7.15) rather than (7.12). Analyses of the data based on this sort of approach have been presented by Collins and Kearney (1983), and by Coon *et al* (1978) who discuss alternative trajectory end points ($\alpha(t) \to -2$ or -3) suggested by different ways of applying the dimensional counting rules. However, we should stress again that both the form of the Reggeisation in QCD and the overlap of the Regge and large-angle limits are plausible speculations, not proven results.

Figure 71 suggests a physical description of the Reggeon exchange process. As the hadrons approach each other the quarks are mutually attracted and slow down by radiating virtual gluons, as in figure 72(a). By absorbing the gluons emitted from the other quark, each quark can reverse its direction of motion, and thus they can be exchanged between the hadrons. If a q and a $\bar{q}$ attract each other, as in figure 72(b), they can annihilate, but subsequently another $q\bar{q}$ pair must be created so that colourless hadrons can appear in the final state. Of course, the likelihood of just two outgoing hadrons mopping up all the virtual gluons in this way is rather small at high energies, which is presumably why the Reggeon exchange cross sections fall with s. Thus the dominant Reggeons (which are shown in figure 9) have $\alpha(t \approx 0) \approx 0.5$ and so $d\sigma/dt \sim s^{-1}$ at small $|t|$ from (7.9). Heavier quarks are obviously harder to 'turn

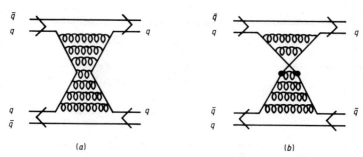

Figure 72. Parton description of Reggeon exchange. In (*b*) the dots show the annihilation and subsequent creation of $q\bar{q}$ pairs which does not necessarily have to occur on the same gluon line.

round' and so the intercepts of the trajectories containing *s, c* or *b* quarks are lower than those in figure 9, giving a more rapidly decreasing cross section. It is more likely that some of the gluons will hadronise independently of the forward-going quarks, as in figure 15, and so multiparticle final states are much more likely than two-particle ones as *s* increases.

This will form the subject of the next chapter.

8

INCLUSIVE REACTIONS

In the early 1970s the study of inclusive processes dominated hadronic-interaction physics. In a typical ISR collision about 18 hadrons are produced and we are clearly forced to be selective about the information obtained and studied. A useful choice is to study inclusive processes of the type

$$A + B \rightarrow C + X$$

where C is the observed particle and X represents everything else produced. Such reactions are described by three independent kinematic variables, for example, s, t of (2.1), (2.10) together with the missing mass squared:

$$M^2 \equiv (p_A + p_B - p_C)^2 \tag{8.1}$$

see figure 73. Alternatively we may use s with p_L, p_T the longitudinal, transverse components of $\boldsymbol{p}_C$. As noted in § 2.1 the p_L dependence is usually expressed in terms of x ($\equiv p_L/p$ in the CM frame) or the rapidity y of (2.9). In the CM frame, $\boldsymbol{p}_A + \boldsymbol{p}_B = 0$, we see from (8.1) that

$$M^2/s \simeq (1-x) \tag{8.2}$$

for $p_L \gg p_T$ and large s. For a fixed s the extremities of the x plot are populated by low missing mass.

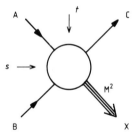

Figure 73. Kinematic variables for $AB \rightarrow CX$ where X represents all the other particles produced.

8.1 Mueller's theorem

A major stimulus for the study of inclusive reactions was the prediction by Feynman (1969) (and also by Yang and collaborators: Benecke *et al* (1969)) that the cross section would scale at high energies, i.e. the cross section would be a function of two variables x and p_T, and not of the energy $\sqrt{s}$ directly. The arguments for 'Feynman' scaling were based mainly on physical intuition. (We have already encountered an analogous, but distinct, scaling prediction by Bjorken for deep inelastic lepton scattering $e + p \rightarrow e + X$.) However, it was soon shown that these results follow from a generalisation of the application of Regge theory to the optical theorem.

To introduce this idea it is useful to recall the ordinary optical theorem (7.16) which relates, via unitarity, the total cross section $(A+B \to X)$ to the imaginary part of the forward elastic scattering amplitude

$$\sigma_T(AB) \simeq \frac{1}{s} \, \text{Im} \, A(AB \to AB)\big|_{t=0}. \tag{8.3}$$

This was represented pictorially in figure 65. The total cross section is very complicated, being the sum of the cross sections of many multiparticle reactions, each with their own energy dependence. Yet, through the magic of the optical theorem, it is given by a single two-body elastic scattering amplitude. Thus the properties of the latter (Regge asymptotic behaviour, duality properties, factorisation, etc) can be used to determine the behaviour of the total cross section as in (7.22).

Mueller (1970) extended the optical theorem to relate the inclusive cross section for $A+B \to C+X$ to an elastic three-body forward amplitude

$$f \equiv E_C \frac{d\sigma}{d^3 p_C} (A+B \to C+X) \simeq \frac{1}{s} \, \text{Disc}_{M^2} \, A(AB\bar{C} \to AB\bar{C}) \tag{8.4}$$

where the discontinuity is to be taken only across the M^2 cut of the elastic amplitude. This is represented pictorially in figure 74. The generalisation is not as straightforward as it looks, however. Since C is an outgoing particle we are not in the physical region of the elastic process $AB\bar{C} \to AB\bar{C}$. We need to make a delicate analytic continuation of the (many-variable) three-body amplitude from the physical region of $A+B \to C+X$.

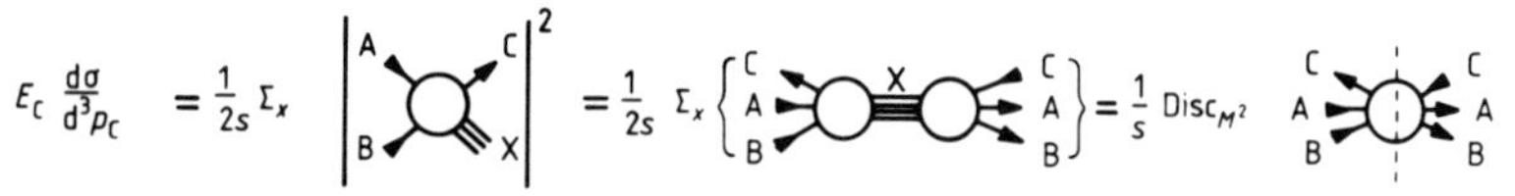

Figure 74. Pictorial representation of Mueller's optical theorem.

(This continuation has been justified for field theoretic models.) The importance of Mueller's optical theorem is that the behaviour of the inclusive reaction can be obtained from that of the, theoretically much simpler, elastic three-body amplitude.

To predict the inclusive cross section for $A+B \to C+X$ we shall assume Regge behaviour for the three-body amplitude. The predictions are therefore relevant to large s and small p_T. It is necessary to distinguish between three kinematical regions for particle C. The beam and target fragmentation regions correspond to finite $t \equiv (p_A - p_C)^2$ and $u \equiv (p_B - p_C)^2$, respectively. Intuitively C can be regarded as a fragment of A (or B) if, as in figure 15, the momentum difference between the two remains finite at large s. The third region is the central region where t and u are both large and C is not closely associated with either incoming particle.

A general criterion for a Regge expansion of a multiparticle amplitude, such as $AB\bar{C} \to AB\bar{C}$, is that the kinematic invariant spanning the Regge exchange should be large (say $\geq 5 \text{ GeV}^2$). In the central region, where t and u are both large, we have the double Regge limit. On the other hand, in the beam fragmentation region ($s \to \infty$, t finite) there are three different Regge limits when different combinations of the invariants M^2 and s/M^2 become large. We discuss these first.

8.2 The fragmentation region

Typical Regge contributions for the three different limits are shown in figure 75. We shall discuss these in turn.

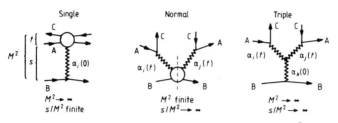

Figure 75. Regge limits in the beam fragmentation region A $\xrightarrow{\text{B}}$ C.

(i) The single Regge limit ($M^2 \to \infty$, s/M^2 fixed). Here the Regge expansion of the forward elastic AB$\bar{\text{C}}$ amplitude in (8.4) gives an inclusive cross section (2.5) of the form

$$f = \sum_i \beta_i \left(\frac{s}{M^2}, t \right) (M^2)^{\alpha_i(0)-1}. \tag{8.5}$$

Assuming the leading singularity, the Pomeron, has $\alpha_P(0) = 1$ and the leading meson exchanges have $\alpha_R(0) = \frac{1}{2}$ we have scaling of the inclusive cross section as $s \to \infty$ and a $s^{-1/2}$ approach to the scaling form.

(ii) The normal Regge limit (M^2 finite, $s/M^2 \to \infty$). In this case (8.4) yields an inclusive cross section:

$$f = \frac{1}{s} \sum_{ij} \beta_{\text{AC}}^i(t) \beta_{\text{AC}}^j(t) s^{\alpha_i(t)+\alpha_j(t)} \, \text{Disc}_{M^2} A(\alpha_i B \to \alpha_j B) \tag{8.6}$$

which contains the Reggeon-particle forward-scattering amplitude with maximal helicity flip of the Reggeon.

(iii) The triple Regge limit ($M^2 \to \infty$, $s/M^2 \to \infty$). This kinematic region is the overlap of the above two regions. For large M^2 we can expand the Reggeon-particle amplitude of (8.6) in terms of Regge exchanges in the B$\bar{\text{B}}$ channel:

$$f = \frac{1}{s} \sum_{ijk} \beta_{ijk}(t) s^{\alpha_i(t)+\alpha_j(t)} (M^2)^{\alpha_k(0)-\alpha_i(t)-\alpha_j(t)}. \tag{8.7}$$

Although applicable only in a very limited kinematic domain, triple Regge analyses of the data have been very successful. Many tests, some involving dual model arguments, have been performed. A recent review has been given by Ganguli and Roy (1980).

As an illustration consider the determination of the ρ trajectory from $\pi^{\pm} p \to \pi^0 X$ data at 100 GeV/c (Barnes *et al* 1978). The dominant triple Regge contribution has $\alpha_i = \alpha_j = \alpha_\rho$ together with a Pomeron, assumed to be $\alpha_k(0) = 1$. Thus, using (8.2), the inclusive cross section is

$$f = \beta_\rho(t)(1-x)^{1-2\alpha_\rho(t)}. \tag{8.8}$$

The x dependence of the data in the triple Regge region ($0.8 < x < 0.98$), for fixed values of t, is found to give a reasonable ρ trajectory, $\alpha_\rho(t)$, for $-t \lesssim 1.5$ (GeV/c)^2.

At higher values of $-t$ the trajectory levels off at about -0.5. This may be attributed to the hard scattering effects discussed in the previous chapter. The trajectories approach negative integers if the Regge exchanges become hard scattering terms as $t \to -\infty$. The idea is that for large angle $AB \to CX$, for example figure 76, only the most elementary constituent interchange (CIM) processes such as $Aq \to Cq$ contribute (Sivers *et al* 1976). Then $\alpha_{AC}(-\infty)$ is determined by the requirement that the dimensional counting rule (6.7) should agree with (8.8), i.e. we need

$$\alpha_{AC}(-\infty) = 1 - n_s$$

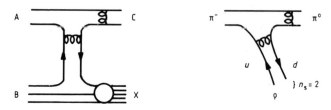

Figure 76. The CIM contribution $Aq \to Cq(\pi^- u \to \pi^0 d)$ to $AB \to CX(\pi^- p \to \pi^0 X)$.

where n_s is the minimum number of effective spectators in the transition $A \to C$. In the low-t region the trajectory is lifted above this value by gluon bremsstrahlung, as described in § 7.3. Thus, for our example $\pi^- p \to \pi^0 X$ the typical CIM process is $\pi^- u \to \pi^0 d$, shown in figure 76(b), and there are two spectators, and two hard gluons are needed for large-angle scattering. Hence we expect $\alpha_\rho(-\infty) = -1$. However, estimates of the magnitude of this CIM contribution are well below the data (Barnes *et al* 1978).

An alternative interpretation of the data is that one should add the triple Regge contribution to the recombination model of chapter 6 (Van Hove 1979), the former dominating for $x \geq 0.8$ and the latter for smaller x.

The triple Regge analysis of diffractive data, particularly $pp \to pX$, allows a study of the, theoretically important, triple Pomeron, PPP, and PPR couplings. There were discrepancies between the original analyses (Field and Fox 1974, Roberts and Roy 1974) which were due to the use of different data compilations. Recent data support the results of the latter analysis and give a non-vanishing triple Pomeron coupling at $t = 0$ and also a dominant triple Pomeron component to diffractive processes, i.e. $G_{PPP} \gg G_{PPR}$ (see Ganguli and Roy 1980). We shall examine this triple Pomeron contribution further in § 10.3.

8.3   The central region

In the central region for $A + B \to C + X$ the invariants s, t, u are all large:

$$t = -\sqrt{s}(E - p_L) \qquad u = -\sqrt{s}(E + p_L). \qquad (8.9)$$

However

$$\kappa^2 \equiv ut/s = E^2 - p_L^2 = m_C^2 + p_T^2 \qquad (8.10)$$

is finite.

In this region the double Regge limit (figure 77) is appropriate for the $A B \bar{C} \rightarrow A B \bar{C}$ amplitude. The inclusive cross section is then of the form

$$f = \sum_{ij} \beta_{ij}(\kappa) |t|^{\alpha_i(0)-1} |u|^{\alpha_j(0)-1}. \tag{8.11}$$

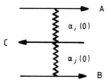

Figure 77. Double Regge limit in the central region.

With the canonical choice $\alpha_P(0) = 1$ and $\alpha_R(0) = \frac{1}{2}$ this becomes

$$f = \beta_{PP}(\kappa) + \kappa^{-1/2} s^{-1/4} \sum_R [\beta_{PR}(\kappa) \exp(-y/2) + \beta_{RP}(\kappa) \exp(y/2)] + O(s^{-1/2})$$

where here y is the centre-of-mass rapidity, see (2.9). Thus asymptotically we expect the inclusive cross section to scale, and also to be independent of y with a central plateau. An $s^{-1/4}$ approach to scaling is predicted. From duality arguments the leading non-scaling terms are found to be positive and so the approach to scaling in the central region should be from above. Experimentally, however, all the central cross sections are found to rise with energy. This was suspected from the beginning, but is now confirmed by recent ISR data, which have shown a rise of as much as 40% in the $pp \rightarrow \pi^- X$ central cross section over the ISR energy range. There have been many models attempting to explain this effect, and also the shape of the y distribution and the shrinkage of the p_T distribution with increasing energy. They are discussed in the review of Ganguli and Roy (1980). Most anticipate that scaling will occur when high enough energies have been achieved.

We should stress that the Mueller–Regge formalism, though extremely powerful for inclusive phenomenology, is not a substitute for a dynamical model for multiparticle production. For instance, it does not explain the nature of the Pomeron singularity. This will be the subject of the next chapter.

9

THE POMERON AND REGGE CUTS

9.1 The Pomeron pole

The approximate constancy with s of all the high-energy forward elastic hadron differential cross sections, and of the hadron scattering total cross sections, implies that the elastic scattering amplitude $A(s, t \sim 0) \sim s$, and hence, if a Regge pole exchange is the dominant mechanism, that there is a trajectory (called the Pomeron, P) with $\alpha_P(0) \simeq 1$ (see (7.8), (7.9) and (7.16)). Since this behaviour seems to be independent of the flavour of the hadrons (and hence of their quark structure) and since the known trajectories involving quark exchange all have $\alpha_R(0) \leqslant \frac{1}{2}$, it is generally supposed that the P represents gluon exchanges.

A physical picture of how this may happen was given by Low (1975) (see also Nussinov 1976). As the colour singlet hadrons approach near each other a colour octet gluon may be exchanged between them (figure 78(a)). But as a result each hadronic cluster then becomes an octet, and as they attempt to move apart the colour lines of force connecting them get stretched. Only by the exchange of another gluon can the clusters become colourless, and hence be free to separate. Thus the fundamental diagram for the elastic scattering amplitude should be two-gluon exchange (figure 78(b)). Since the gluon has spin 1, the single-gluon exchange amplitude $\sim s$ at fixed t (see table 3) and the two-gluon box like $\sim (i/s)ss = is$ (the i/s arising from the integration over intermediate states) which is just the phase and energy dependence of (7.11) with $\alpha = 1$ and even signature ($\mathcal{S} = +1$). (We assume that the IR divergences of massless gluon exchange cancel for composite systems.)

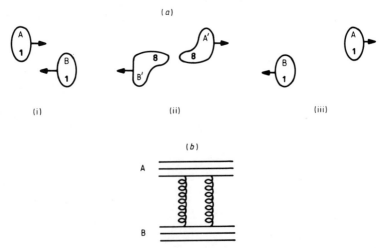

Figure 78. Two-gluon exchange in AB elastic scattering. After single-gluon exchange the colour singlets A, B become colour octets A′, B′, and a further gluon exchange is necessary to produce outgoing singlets.

Inserting gluons and/or $q\bar{q}$ pairs to make ladders like figure 79 may be expected to produce a Pomeron trajectory, perhaps with $\alpha(0) > 1$, somewhat like the mechanism discussed in § 7.3. Indeed the slow rise of σ_T with s at the highest energies in figure 10 can be explained if $\alpha_P(0) \simeq 1.07$ (Collins *et al* 1974). Such a rising power behaviour should not continue indefinitely, however (Froissart 1961, Martin 1963, 1966). Instead we expect that the simultaneous exchange of many Pomeron ladders will change the asymptotic behaviour to at most a $\log^2 s$ behaviour, as required by unitarity (see Collins 1977, p 278 for the details).

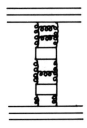

Figure 79. Gluon ladders (with possible quark loop insertions) which may sum to produce a Pomeron trajectory.

The finite range of the hadronic interaction, R (given by the exchange of the lightest hadron, the pion—see § 1.5), means that only in low angular momentum states l such that $l \leqslant l_{max} \simeq R\sqrt{s} \log s$ can scattering occur. If the hadrons pass each other at higher l their impact parameter is greater than R and so they miss each other. Partial-wave unitarity (i.e. conservation of probability) demands that each partial-wave amplitude obeys Im $A_l(s) \leqslant 1$. Hence from the optical theorem (7.16)

$$\sigma_T(s) = \frac{1}{s} \operatorname{Im} A(s, 0) = \frac{1}{s} \sum_{l=0}^{\infty} (2l + 1) \operatorname{Im} A_l(s)$$

$$\leqslant \frac{1}{s} \sum_{l=0}^{l_{max}} (2l + 1) \leqslant R^2 \log^2 s \simeq \frac{1}{m_\pi^2} \log^2 s. \tag{9.1}$$

Thus the total cross section can grow no more rapidly with s than $\log^2 s$, and not like $s^{\alpha_P(0)-1}$ with $\alpha_P(0) > 1$. This is called the 'Froissart bound'. However, even at ISR energies where $\log s \simeq 8$ we find that $\sigma_T(pp)$ is only of the order of $1/m_\pi^2$ (figure 10) and hence the cross section is still about two orders of magnitude below the Froissart bound. Only at astronomically high energies does the observed $s^{0.07}$ behaviour exceed (9.1), so it is not clear that the mechanisms which will ultimately ensure satisfaction of the Froissart bound are operative at attainable energies. Other explanations for the rise of $\sigma_T(s)$ involving multiple P exchanges are discussed below. But there is good experimental evidence that the Pomeron coupling factorises (see, for example, Cool *et al* 1981) as one would expect for a pole, but not necessarily for a cut. However, the precise nature of the Pomeron is still obscure and although we shall generally refer to it as a pole it is important to keep in mind that it may be some much more complex object, which results from the diffractive nature of elastic scattering, and which is only simulating the properties of a pole at currently available energies.

We have seen that quark exchange ladders (figure 70) result in Regge trajectories. The particles on these trajectories arise through the binding of the quarks. Figure

79 might similarly be interpreted as the exchange of flavourless particles made just from gluons. The existence of such 'glueballs' (Jaffe and Johnson 1975) has often been conjectured, but so far there is no definitive proof that any observed particle is a glueball rather than an ordinary $q\bar{q}$ state (Close 1983). It is, of course, perfectly possible that some of the known flavourless particles are mixtures of glueballs and $q\bar{q}$ resonances, exemplified by the insertion of quark boxes in figure 79. However, the success of Zweig's rules (see § 7.2) suggests that such mixing may be small for high-mass states, and the occurrence of a spin-2 glueball at a mass m_g such that $\alpha_P(m_g^2) = 2$ cannot be ruled out. Such a weakly-coupled meson is very difficult to isolate, particularly since the small slope of the Pomeron trajectory ($\alpha'_P \approx 0.25\,\mathrm{GeV}^{-2}$) suggests $m_g \approx 2\,\mathrm{GeV}/c^2$.

The final-state hadrons of figure 78 do not have to be identical to the incoming ones. The gluon exchanges may result in excitation of the quarks, so that the outgoing particle has the same flavour content, but higher mass and possibly angular momentum. This is known as 'diffractive excitation'. Examples are $\pi p \to \pi N^*$ (where N^* is an excited $I = \frac{1}{2}$ nucleon state) and $\pi p \to A_2 p$, since the A_2 has the same quark content as π but higher spin (see table 2). In all such cases it is found that at high energy the cross section is essentially independent of energy, unlike the flavour exchange processes discussed in chapter 7. Likewise in the Mueller–Regge approach to inclusive processes of chapter 8 we have observed the need for P exchange in the elastic amplitude for $AB\bar{C} \to AB\bar{C}$.

Naively one might regard figure 79 as qq scattering, the remaining quarks in the hadrons being simply spectators. If so, for particles containing identical quarks, one has (Kokkedee 1969) $\sigma_T(MM) = 4\sigma_T(qq)$, $\sigma_T(MB) = 6\sigma_T(qq)$ and $\sigma_T(BB) = 9\sigma_T(qq)$, the coefficients being the number of ways of pairing the valence quarks. Hence, one predicts that, for example, $\sigma_T(pp)/\sigma_T(\pi p) = 1.5$ (since π and p are made of identical u and d quarks) in fair agreement with the ratio 1.7 found at the highest energies in figure 10. From this viewpoint one might also expect that the pp elastic differential cross section would fall more rapidly with increasing $-t$ than that for πp, in the ratio $(F_B(t)/F_M(t))^2 \sim t^{-2}$, because of the greater difficulty of reforming the $3q$ particle in the final state, as discussed in § 2.3. This is also in accord with experiment (Collins and Wright 1979).

Further confirmation that P exchange often involves essentially just a single quark from each hadron, as in figure 79, comes from the success of the hypothesis of f dominance of the P coupling (Carlitz et al 1971). The Pomeron has the same flavour properties as the f meson and hence one may expect that the longest-range part of the coupling at small $|t|$ will stem from the formation of a virtual f meson as in figure 80. (This is analogous to the vector dominance hypothesis that the photon couples to

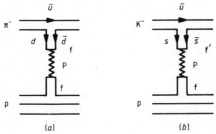

Figure 80. f(f') dominance of the Pomeron coupling in $\pi^- p$ and $K^- p$ elastic scattering.

a hadron via the lightest available vector mesons—see § 2.3 and (2.20).) Hence the P-exchange amplitude involves the inverse propagator $(\alpha_P(t) - \alpha_f(t))^{-1}$ representing the amount by which the f is off its 'mass shell' in angular momentum if it is to couple to the P. For example, in πp both the valence quarks of the π are capable of forming an f, so there are two diagrams like figure $80(a)$. But in Kp scattering only one of the K quarks can form an f, the other giving rise to the heavier $f'(s\bar{s})$ state as in figure $80(b)$ (see table 2) which takes the coupling further off its mass shell. It is thus predicted that at high energies

$$\frac{\sigma_T(Kp)}{\sigma_T(\pi p)} = \frac{(\alpha_P(0) - \alpha_f(0))^{-1} + (\alpha_P(0) - \alpha_{f'}(0))^{-1}}{2/(\alpha_P(0) - \alpha_f(0))} \approx 0.8 \tag{9.2}$$

in good agreement with figure 10.

9.2 Regge cuts

It is possible for more than one Reggeon to be exchanged, as in figure $81(a)$. It will be noted that the two Reggeons are coupled to different constituents of the hadron. This is because at high energies the hadrons pass each other very rapidly and the chance of the same pair of constituents interacting twice (as in figure $81(b)$) falls rapidly with s. This is verified in field theory models in which the Reggeons are represented by sums of ladders, as discussed in chapter 7. It can be shown that the exchange of two or more Reggeons will give rise to a branch cut in the t-channel angular momentum plane, at $l = \alpha_c(t)$, called a Regge cut (see Collins (1977 chap 8) for a detailed discussion).

Schematically, a two-Reggeon cut amplitude, like $R_1 \otimes R_2$ of figure $81(a)$, is given by

$$A^c(s, t) \sim \frac{i}{|s|} \int d\Phi_{12} A^{R_1}(s, t_1) A^{R_2}(s, t_2) \tag{9.3}$$

where the integral is over the two-Reggeon phase space (see equation (8.4.1) of Collins (1977) for a more exact expression). The position of the branch point trajectory is given by

$$\alpha_c(t) = \max \{\alpha_1(t_1) + \alpha_2(t_2) - 1\} \tag{9.4}$$

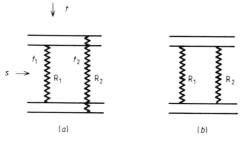

Figure 81. (a) Two-Reggeon exchange, denoted $R_1 \otimes R_2$, with momentum transfers given by t_1, t_2, respectively, gives rise to a cut in the complex l plane. Planar diagram (b) gives a negligible contribution at high s.

the maximum being taken over the available phase space, and the asymptotic form of the cut amplitude is

$$A^c(s, t) \underset{s\to\infty}{\sim} \frac{s^{\alpha_c(t)}}{\log s} \tag{9.5}$$

the log s arising from the integration.

Cuts involving P exchange have particularly interesting properties. For a linear P trajectory, $\alpha_P(t) = 1 + \alpha'_P t$, (9.4) gives a $P \otimes P$ cut at

$$\alpha_c(t) = 1 + \frac{\alpha'_P}{2} t \tag{9.6}$$

so that the pole and cut coincide at $t = 0$, and the cut lies above the pole for $t < 0$. Since the P amplitude is almost pure imaginary for $t \approx 0$, $A^P \sim is$, (9.3) gives a cut amplitude of the form $(i/s)is\ is/\log s = -is/\log s$, i.e. modulo log s, it has the same energy dependence as the pole but the opposite phase. It may be anticipated that the P pole will dominate at small $|t|$, but that the cut with its flatter t dependence will take over at larger $|t|$, and that there will be an intermediate region where the amplitude is very small because of the destructive interference between the pole and the cut, with their opposed phases. The pp (and p̄p) differential cross sections exhibit a sharp dip at $|t| = 1.4 \text{ GeV}^2$ (figure 82) which is readily interpreted in this way (Collins and

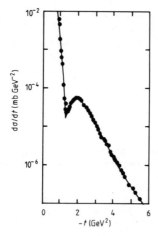

Figure 82. The differential cross section for pp elastic scattering at $\sqrt{s} = 53$ GeV measured at the ISR. The fit is from Collins and Gault (1978).

Gault 1978). If $\alpha_P(0)$ is above 1 the cuts will also be above 1 and the asymptotic behaviour is controlled by a superposition of multi-P cuts which obeys the bound (9.1) (see, for example, White 1980). A similar destructive interference between the ρ pole and a $\rho \otimes P$ cut can be used as an alternative explanation for the dip at $|t| = 0.5 \text{ GeV}^2$ in figure 63 to the nonsense decoupling mechanism discussed in § 7.1. A comprehensive analysis of flavour-exchange processes with Regge cuts along these lines has been presented by Kane and Seidl (1976).

Equation (9.3) may be generalised to include the exchange of any number of Reggeons (see, for example, Collins 1977 chap 8) and it is found that the more

Reggeons are exchanged the flatter is the branch point trajectory as a function of t. So from the Regge theory standpoint scattering processes seem to become more and more complicated as $-t$ increases; with the dominance of multi-Reggeon exchange. This is to be contrasted with the parton viewpoint, that only simple parton diagrams (like, for example, figure 71(a)) determine the scattering process at large t.

The resolution of this conflict is straightforward if the Regge trajectories depart from linearity at large $|t|$ and asymptote to negative integers (probably -1 for the dominant exchange?) as discussed in § 7.3. In this case (9.4) gives

$$\alpha_c(-\infty) = \alpha_1(-\infty) + \alpha_2(-\infty) - 1$$

and so the cut ends up below the poles. We are thus left with the possibility that Regge poles dominate at small $|t|$, and, through the bending of the trajectories, as $t \to -\infty$, with an intermediate $|t|$ region (typically 0.5–$2\,\text{GeV}^2$) where cuts are also important (see Collins and Wilkie 1981). This seems much more attractive than the supposition that the large-angle hard scattering mechanisms are quite distinct from the small-angle Regge exchange dynamics, with no simple way of interpolating between them. Detailed parametrisations based on these hypotheses have been quite successful (Collins and Kearney 1983).

9.3 Trajectory dynamics and particle production

In figures 71 and 79 we have drawn ladder diagrams to suggest how Reggeons may be built up in QCD from multi-parton intermediate states in the s channel. In fact, of course, since the partons are confined such diagrams do not represent possible physical states, and hence cannot properly display the unitarity structure of the theory. Instead we should include only multiparticle intermediate states as in figure 83(a) for which the QCD interpretation is something like figure 83(b). Only that subset of parton states in which the coloured partons are gathered into colourless hadronic clusters are to be allowed as intermediate states in these diagrams (Nussinov 1976). The optical theorem equates the sum of multiparticle cross sections for processes like figure 84 (the total cross section) to the imaginary part of the elastic amplitude of figure 83(b) (cf figure 65).

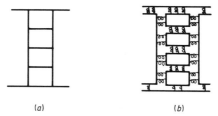

(a) (b)

Figure 83. Ladder diagrams which sum to give Regge behaviour. In (a) we show the physical particle intermediate states, and in (b) we symbolically sketch their parton composition.

The central region of figure 84 (x near 0) in which a quark from one particle and an antiquark from the other attract each other, slow down by the bremsstrahlung of gluons and hence ultimately hadrons, and finally annihilate each other, looks just like the inverse of figure 41(c) in which the $q\bar{q}$ pair separate and lose energy radiating hadrons. We might thus expect that the average particle multiplicity in hadron scattering $\langle n \rangle_h$ would be very similar to that in e^+e^- annihilation $\langle n \rangle_{e^+e^-}$ at the same energy, if allowance is made for the fact that some of the energy is carried by the spectator quarks (the top and bottom lines of figure 84) which populate the fragmentation regions, $|x|$ near 1, and which we discussed in § 6.1. In particular, for collisions involving protons the spectator system is a diquark which may produce a jet with a quite different multiplicity structure to those of the quarks or gluons (see, for example, Gunion 1980). An attempt has been made to allow for these effects (Basile *et al* 1980, 1981) by taking the energy available for hadron production to be

$$E_{\text{hadron}} = E_{\text{beam}} - E_{\text{fastest proton}}$$

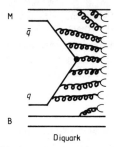

Figure 84. A multiparticle production process.

and comparing the pp multiplicity at this lower energy with that in e^+e^-. This shift of the pp curve in figure 11(a) results in quite good agreement with the e^+e^- multiplicity.

However, the dominant mechanism in high-energy hadron scattering is presumably not quark exchange (which gives the normal Regge trajectories with $\alpha_R(0) \leq \frac{1}{2}$) but gluon exchange, which we have identified with the Pomeron. If the P couples through the f as in figure 80, this need not necessarily make much difference (see figure 85(a)). A simple physical interpretation of the diagram is that it is the slow sea quarks from the hadrons which interact via the exchange of two (or more) gluons. But the Low–Nussinov mechanism of gluon exchange followed by hadronisation in the colour field produced by the separating colour octets suggests bremsstrahlung through a gluon jet, as in figure 85(b). Naively this would lead one to expect that $\langle n \rangle_h = \frac{9}{4} \langle n \rangle_{e^+e^-}$ due

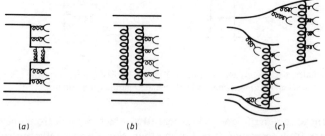

Figure 85. Multiparticle production by Pomeron exchange assuming (a) f dominance; (b) hadronisation direct from gluon exchange; (c) a valence quark from each hadron is 'held back' so that two independent chains of hadrons are formed.

to the fact that the gluon–gluon vertex coupling is stronger than the quark–gluon vertex (Brodsky and Gunion 1976).

Alternatively, if the 'held-back' effect discussed in § 6.1, in which one valence quark from each hadron interacts with (is held back by) the remaining quarks of the other hadron, is correct, then two independent chains of hadrons are produced, as in figure 85(c). These may be quark chains, gluon chains or some mixture of the two, as shown. Via the optical theorem (cf figure 65) these two chains produce the cylinder of figure 54, and give the Pomeron its distinctive topological structure expected in dual models. This leads one to expect that $\langle n \rangle_h \gtrsim 2\langle n \rangle_{e^+e^-}$ (Capella 1981), contrary to the data of figure 11(a). However, consistency with these data is still possible if allowance is made for the diquark fragmentation and the fact the two chains have to share the incident energy. Only at very high energies when central region production dominates should $\langle n \rangle_{pp} \rightarrow 2\langle n \rangle_{e^+e^-}$. But as spin-1 gluon exchange is not the dominant mechanism here it is not obvious that the total cross section must be constant with s. A detailed discussion of these approaches has been given by Gunion (1980) (see also Capella 1981).

Clearly much remains to be understood about the precise mechanisms of particle production which build up the total cross section. As mentioned in chapter 5 we still do not have definite experimental evidence that gluons are much more effective than quarks at radiating hadrons and so produce jets of higher multiplicity, as the naive colour coupling arguments suggest. Even more important, there is no obvious reason why these soft scattering processes should have any simple parton description. Since α_s is of the order of 1, parton interactions of arbitrary complexity all seem equally probable. However, the approximate scale invariance of hadron cross sections, and the similarity of the hadron multiplicities, etc, in forward scattering, and in hard processes like large p_T jets, deep inelastic scattering and e^+e^- annihilation, permits some optimism that a better understanding of the Pomeron in terms of QCD may emerge.

10

HADRON SCATTERING AT COLLIDER ENERGIES

We have remarked several times that the parton structure of hadrons is most clearly revealed in very high-energy collisions, and so the best data with which to test the parton model should come from the highest energy hadron scattering facilities. At the time of writing (1983) these are the intersecting storage rings (ISR) and the antiproton–proton collider (which we shall refer to as the Collider) at CERN. In the former, two beams of protons, and more recently beams of protons and antiprotons, with energies up to 31.5 GeV in each beam collide to give a maximum $s = (2 \times 31.5 \text{ GeV})^2 = 3969 \text{ GeV}^2$ (which is equivalent to $p_{\text{lab}} \approx 2100 \text{ GeV}/c$ for collisions on a stationary proton target). At the Collider, 270 GeV beams of protons and antiprotons produced by the super-proton-synchrotron (SPS) are collided, providing $s = (2 \times 270 \text{ GeV})^2 \approx 2.9 \times 10^5 \text{ GeV}$. These s values greatly exceed those available for any other hadron scattering processes, such as πp or Kp for example, where a secondary meson beam is incident upon a stationary proton target. Although the collision rate in these fixed-target laboratories is much greater, the highest s available (at the CERN SPS and at the Fermi National Laboratory in the USA) is less than 1000 GeV^2. In the following chapters we shall therefore concentrate on the information about the nature of hadrons which has been gained at the colliding beam facilities.

10.1 The total cross sections

In figure 86(a) we compare data on the total cross sections for pp and p̄p scattering. It will be noted that whereas $\sigma_T(\bar{p}p) > \sigma_T(pp)$ at low energies they become almost equal at very high s. Figure 86(b) shows how the difference

$$\Delta \sigma \equiv \sigma_T(\bar{p}p) - \sigma_T(pp) \tag{10.1}$$

decreases as a simple power of s, approximately $\sim s^{-0.58}$.

As we discussed in § 7.1 the total cross section is determined through the optical theorem by the forward ($t = 0$) elastic scattering amplitude:

$$\sigma_T(pp) = \frac{1}{s} \text{Im } A(pp \to pp). \tag{10.2}$$

The flavour-dependent part of the elastic scattering amplitude, which accounts for $\Delta \sigma$, is due to the exchange of meson Regge trajectories, as illustrated by the quark line diagrams of figure 87. Only diagram (a) is permitted in pp scattering since the flavour of the proton is carried by the quarks, while only diagram (b) occurs in p̄p scattering. Now according to (7.12) the amplitude of diagram (a) has an asymptotic behaviour at fixed t of the form $\sim (-u)^{\alpha(t)} \to s^{\alpha(t)}$ for $s \gg t$, m_p^2 (see (2.12)). This is purely real

92

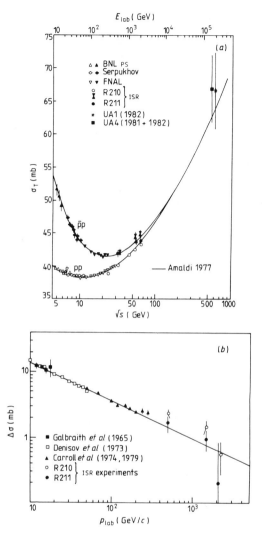

Figure 86. (a) A compilation of the pp and p̄p total cross section data by Matthiae (1983). The full curves are a $\log^2 s$ dispersion relation description by Amaldi *et al* (1977). (b) The difference of the total cross sections $\Delta\sigma \equiv \sigma_T(\bar{p}p) - \sigma_T(pp)$ versus p_{lab} from Matthiae (1983). The straight line is the prediction of ω exchange with $\alpha_\omega(0) \simeq 0.42$.

and so does not contribute to the total cross section (10.2). However diagram (b) behaves like

$$(-s)^{\alpha(t)} \to e^{-i\pi\alpha(t)} s^{\alpha(t)}$$

and so has an imaginary part proportional to $[\sin \pi\alpha(t)]s^{\alpha(t)}$. (Note that all the other factors in (7.11) are purely real.) Hence we expect that $\sigma_T(\bar{p}p)$ will have a contribution $\sim s^{\alpha(0)-1}$ due to $q\bar{q}$ exchange which is absent from $\sigma_T(pp)$, see (7.22). The observed power of $\Delta\sigma$ gives $\alpha(0) = 0.42$.

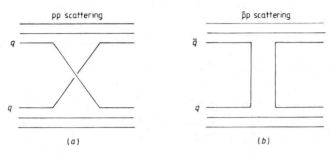

Figure 87. Quark exchanges in (a) pp and (b) p̄p scattering.

Since we find that $\sigma_T(pp) \simeq \sigma_T(np)$ this exchanged trajectory must be made of $u\bar{u}$ and $d\bar{d}$ quarks in an isospin zero combination

$$\frac{1}{\sqrt{2}}(u\bar{u} + d\bar{d})$$

which will couple equally to neutrons and protons (**see table 2**). The leading trajectories containing these quarks are the degenerate f and ω trajectories, and, as will be seen from (7.22) the difference between pp and p̄p can be accounted for by ω exchange which, having odd charge conjugation, contributes oppositely in the two processes. In fact since the exchange amplitudes obey $\mathrm{Im}\,f \simeq \mathrm{Im}\,\omega$ we have (neglecting ρ and A_2 which have isospin one and are small)

$$\sigma_T(pp) \simeq P \sim As^{\alpha_P(0)-1}$$

$$\sigma_T(\bar{p}p) \simeq P + 2\omega \sim As^{\alpha_P(0)-1} + Bs^{\alpha_\omega(0)-1} \qquad (10.3)$$

as in (7.22). Figure 9 shows that indeed $\alpha_\omega(0) \simeq 0.42$, as the behaviour of $\Delta\sigma$ has led us to anticipate, so there can be little doubt that the difference of the cross sections is accounted for by ω meson exchange. The cancellation between f and ω is not exact, however, and this leaves a very small $s^{\alpha_\omega(0)-1}$ contribution which produces the small fall of $\sigma_T(pp)$ observed at low energies in figure 86(a).

But essentially all the pp cross section, and p̄p at very high energies is associated with the Pomeron, P, in (10.3), which we have identified with flavourless gluon exchanges. There are various ways in which one can parametrise this Pomeron contribution. One, which gives rather a good description of the data, is

$$\sigma_T(pp) = 38.3 + 0.43 \log^2\left(\frac{s}{100\,\text{GeV}^2}\right) \text{ mb.} \qquad (10.4)$$

This $\log^2 s$ behaviour is often ascribed to the saturation of the partial waves ($\mathrm{Im}\,A_l = 1$) leading to the Froissart bound (9.1). But $m_\pi^{-2} = 20$ mb in (9.1) is many times the coefficient in (10.4) so it is clear that most high partial waves are very far from being saturated ($\mathrm{Im}\,A_l \ll 1$) at these energies. Probably the simplest assumption which is compatible with the available data is that the Pomeron is effectively just a Regge pole as in (10.3) with $\alpha_P(0) = 1 + \epsilon$ ($\epsilon > 0$) so that $\sigma_T(pp) = A(s/1\,\text{GeV}^2)^\epsilon$. Taking $A \simeq 23$ mb and $\epsilon \simeq 0.08$ gives a satisfactory account of the rise of the total cross section (Collins and Gault 1978, Donnachie and Landshoff 1983). It is interesting that the proton–air-nucleus cross section determined from cosmic ray data in the very high-energy range

10^7–10^9 GeV, which is quite inaccessible to accelerator experiments, shows the same sort of rise with $\epsilon = 0.06 \pm 0.01$ (Hara *et al* 1983).

Models in which the Pomeron exchange is treated like a field, and multiple Pomeron exchanges are then summed, predict $\sigma_T \sim (\log s)^\nu$ with $\nu \approx 0.26$ (Abarbanel *et al* 1975, Moshe 1978, White 1981, 1983). This rise seems too slow to be compatible with the Collider data unless it is supposed that pre-asymptotic effects are still very important even at the highest energies achieved to date.

While the optical theorem, (10.2), determines just the imaginary part of the scattering amplitude, the elastic differential cross section depends on the square modulus of the amplitude

$$\frac{\mathrm{d}\sigma}{\mathrm{d}t}(s, t) = \frac{1}{16\pi s^2}|A(s, t)|^2 = \frac{1}{16\pi s^2}(\mathrm{Re}\, A(s, t)^2 + \mathrm{Im}\, A(s, t)^2) \qquad (10.5)$$

so if we introduce

$$\rho(s) \equiv \frac{\mathrm{Re}\, A(s, 0)}{\mathrm{Im}\, A(s, 0)} \qquad (10.6)$$

we have

$$\frac{\mathrm{d}\sigma}{\mathrm{d}t}(s, 0) = \frac{1}{16\pi s^2}[\mathrm{Im}\, A(s, 0)]^2[1 + \rho(s)^2] = \frac{[1 + \rho(s)^2]}{16\pi}[\sigma_T(s)]^2. \qquad (10.7)$$

Hence a simultaneous measurement of $\sigma_T(s)$ and $\mathrm{d}\sigma/\mathrm{d}t(s, 0)$ permits the determination of $\rho(s)$ and hence the phase of the elastic scattering amplitude. In practice one has to measure $\mathrm{d}\sigma/\mathrm{d}t$ for small $-t$ (i.e. very small angle elastic scattering) and extrapolate the resulting data back to the forward direction, $t = 0$. This must be done with care because the sharp Coulomb scattering peak at very small angles must be subtracted out.

The results are shown in figure 88. We see that $\rho_{pp}(s)$ is negative at low energies but rises to positive values at high energy, while $\rho_{\bar{p}p}(s) \approx 0$ at low energy and rises to meet ρ_{pp} at higher s. We can understand these results in the following way. At high energies we have found that effectively $A(s, t = 0) \sim s^{\alpha_P(0)}$ with $\alpha_P(0) = 1 + \epsilon$ and

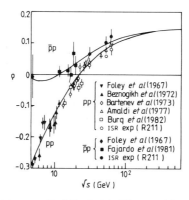

Figure 88. A compilation of data on $\rho = \mathrm{Re}\, A/\mathrm{Im}\, A$ at $t = 0$ for pp and $\bar{p}$p scattering from Matthiae (1983). The full curves are the dispersion relation predictions of Amaldi *et al* (1977).

$\epsilon \simeq 0.08$. Now from (7.13) this power behaviour implies that the phase is

$$\rho(s \to \infty) = -\frac{\cos \pi \alpha(0) + 1}{\sin \pi \alpha(0)} \simeq \frac{\pi \epsilon}{2} \simeq 0.126 \qquad (10.8)$$

since the Pomeron has signature $\mathscr{S} = +1$. At lower energies, where the cross section is falling, effectively $\alpha(0) < 1$ and so $\rho < 0$. In low-energy $\bar{p}p$ scattering there is an additional large contribution from ω exchange with $\alpha_\omega(0) \simeq 0.42$ and $\mathscr{S} = -1$ providing a positive contribution to ρ which dies away with energy relative to the Pomeron but serves to ensure that $\rho_{\bar{p}p}$ is positive unlike ρ_{pp}. The observed phases are thus entirely consistent with the Regge pole phases of (7.11).

Unfortunately these results do not really prove that the amplitudes have to be represented by Regge poles because dispersion relations, which interrelate the imaginary and real parts of the scattering amplitude, require that as long as $\text{Im } A \sim s^\alpha$ then $\rho(s)$ is necessarily given by (7.13) (see Collins 1977, p 194). At the very least, however, the data do demonstrate the utility of representing forward scattering amplitudes by Regge-like power behaviours.

The correct prediction of the power behaviour of $\Delta\sigma$ by the ω trajectory of figure 9, and the approximate equality of $\sigma_T(pp)$ and $\sigma_T(\bar{p}p)$ at high energies helps to justify our assumption that the low-energy, flavour-dependent, part of the cross section is due to Reggeon (i.e. valence quark) exchange, and that the asymptotic part of the cross section is flavour-independent and so presumably due to gluon exchange. Whatever the detailed form of this part of the cross section, whether it is a pole-like (10.3) or more complex multiple exchanges leading perhaps to (10.4), it is convenient to continue to refer to it as the Pomeron.

10.2   The elastic differential cross sections

In figure 89 we illustrate the differential cross section for high-energy pp and $\bar{p}p$ scattering. It will be noticed that, as we anticipated in § 2.2, the scattering is predominantly forward, the cross section being sharply peaked at $t = 0$, the probability of large-angle elastic scattering being very small. In fact the cross section falls roughly exponentially with $-t$, i.e.

$$\frac{d\sigma}{dt} \sim e^{bt} \qquad (10.9)$$

with b about 12 GeV^{-2}. A more detailed examination shows, however, that the rate of fall is greater at very small $|t|$, $0 \le |t| \le 0.2 \text{ GeV}^2$, than for larger values, $0.2 < |t| < 0.8 \text{ GeV}^2$, and that it varies slowly (logarithmically) with energy. At high energies the differential cross sections for pp and $\bar{p}p$ are almost identical, consistent with the idea of flavourless Pomeron exchange.

The change of b with energy is readily understood from the Regge pole viewpoint. The exponential behaviour of the cross section suggests that we should write $F(t)$ in (7.10) as an exponential, say $F(t) = F e^{at}$, in which case (7.10) becomes

$$\frac{d\sigma}{dt} = F \left(\frac{s}{s_0}\right)^{2\alpha_0 - 2} \exp\{[a + 2\alpha' \log(s/s_0)]t\} \qquad (10.10)$$

so that

$$b = a + 2\alpha' \log s \qquad (10.11)$$

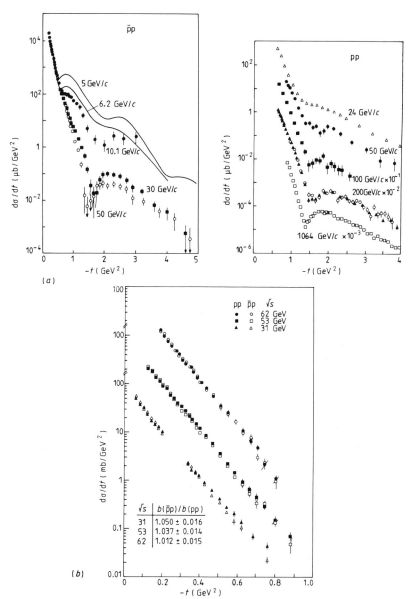

Figure 89. (*a*) Compilation of $\bar{p}p$ and pp elastic scattering data from Asa'd *et al* (1983). (*b*) Data on $\bar{p}p$ and pp elastic scattering at ISR, Breakstone *et al* (1983).

if we set the scale factor $s_0 = 1 \, \mathrm{GeV}^2$. Thus the forward peak sharpens ('shrinks') as $\log s$ increases. Of course at lower energies where the other Reggeons, mainly f and ω, also contribute the analysis is more complicated and $b_{pp} \neq b_{\bar{p}p}$. Figure 90 shows that $b_{pp} \simeq b_{\bar{p}p} \sim \log s$ at large s. The asymptotic slope of b against $\log s$ in figure 90 gives, through (10.11), the slope of the dominant Pomeron trajectory, i.e. $\alpha'_P \simeq 0.1$–0.2, though the recent determination of $b_{\bar{p}p}$ at $\sqrt{s} = 540\,\text{GeV}$ by the UA1 and UA4 collaborations at the $\bar{p}p$ Collider suggests a slightly larger value of α'_P ($\simeq 0.25$). The slope change at $|t| \simeq 0.2\,\text{GeV}^2$ is easily parametrised by rewriting $F(t) = F\,e^{at}[(1-x)+x\,e^{a_1 t}]$ with $a_1 \simeq 3\,\text{GeV}^{-2}$, $x = 0.88$ (Collins and Gault 1978), but no completely convincing explanation for this structure has been proposed. The very small $|t|$ slope found at the Collider, $b = 17 \pm 1\,\text{GeV}^{-2}$ is perhaps a bit larger than might have been anticipated from lower-energy data but generally the forward collider measurements contain few surprises. A good general discussion of the interpretation of these data has been given by Martin (1983).

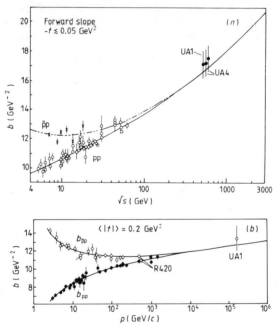

Figure 90. (a) Forward slope parameter b for $\bar{p}p$ and pp scattering, and (b) slope parameter b at $\langle |t| \rangle = 0.2\,\text{GeV}^2$ from Matthiae (1983). The lines are simply to guide the eye.

At lower energies the total cross section, and hence through (10.2) and (10.5) $d\sigma/dt$ at $t = 0$, is larger for $\bar{p}p$ than for pp, but, as figure 90 shows, the slope in t of the $\bar{p}p$ data is also greater and so in the region $|t| = 0.2-0.3\,\text{GeV}^2$ the two cross sections cross over. This effect is just about visible even in the ISR data of figure 89. This difference is due mainly to ω exchange. If we write for the main contributions to the scattering amplitude (cf (7.22))

$$\frac{d\sigma}{dt}(\bar{p}p, pp) = \frac{1}{16\pi s^2}\,|P + f \pm \omega|^2$$

$$\simeq \frac{1}{16\pi s^2}\{|P|^2 + 2\,\text{Re}[P\cdot(f\pm\omega)]\} \qquad (10.12)$$

since $|P| \gg |f|, |\omega|$ and P is predominantly imaginary as $\alpha_P \simeq 1$ then it is evident that the difference between the two differential cross sections is given approximately by $4 \operatorname{Im} P \cdot \operatorname{Im} \omega$. The equality of the two differential cross sections implies that $\operatorname{Im} \omega = 0$ at the cross-over point $|t| = 0.2$–$0.3 \, \mathrm{GeV}^2$. This cross-over effect is also seen in other ω exchange reactions and is probably due to the interference of the ω pole with an $\omega \otimes P$ exchange cut as in (9.3) (see Collins 1977, Kane and Seidl 1976, Irving and Worden 1977). The most important point, however, is that ω exchange seems to account rather well for the pre-asymptotic differences between $\bar{p}p$ and pp scattering.

The most prominent feature of figure 89 is the sharp dip in the differential cross section at $|t| \simeq 1.4 \, \mathrm{GeV}^2$ in both pp and $\bar{p}p$. Closer examination shows that this dip, which is present as a shoulder even at quite low energies, moves in slowly towards smaller values of $|t|$ as s increases. In $\bar{p}p$ scattering the dip is already well established at $s \simeq 60 \, \mathrm{GeV}^2$ but in pp scattering it is not until $s \simeq 200 \, \mathrm{GeV}^2$ that the shoulder seen at lower energies has turned into a really well defined dip. In pp this dip is still present at the highest energy at which elastic scattering has been measured ($s \simeq 3900 \, \mathrm{GeV}^2$) while in $\bar{p}p$ the dip is certainly present up to $400 \, \mathrm{GeV}^2$ but unfortunately there are as yet no measurements for this process in the ISR energy range at sufficiently large values of $|t|$. For $s > 200 \, \mathrm{GeV}^2$ pp and $\bar{p}p$ seem to be essentially identical and their cross sections vary only logarithmically with energy.

All of this is entirely consistent with Pomeron dominance but the dip requires that there be a cancellation between two almost energy-independent contributions which have essentially the same phase but rather different t dependences. The most obvious explanation is that there is an interference between single scattering, represented by the Pomeron pole trajectory (figure 79) and double scattering represented by the exchange of two Pomerons as in figure 81(a). In § 9.2 we discussed how if the P pole amplitude is essentially purely imaginary in its phase then the simultaneous exchange of two P's would give rise to a Regge cut whose amplitude has this same imaginary phase but the opposite sign. Thus if we approximate the P amplitude by

$$A^{P}(s, t) = \mathrm{i} s A_1 \exp(a_1 t) \tag{10.13}$$

(which follows from (7.11) with $\alpha_P(t) = 1$, $\mathscr{S} = +1$, and a residue with an exponential t-dependence) then the two-Pomeron cut evaluated from (9.3) will take the form

$$A^{PP}(s, t) = \frac{-\mathrm{i} s A_2 \exp(a_2 t)}{\log s} \tag{10.14}$$

with $A_2 < A_1$ and $a_2 < a_1$. So the sum

$$A^{P} + A^{PP} = \mathrm{i} s A_1 \, \mathrm{e}^{a_1 t} \left(1 - \frac{A_2}{A_1} \frac{\exp(a_2 - a_1)t}{\log s} \right) \tag{10.15}$$

will have a zero in $|t|$ whose position moves in towards $t = 0$ as $\log s$ increases, and the resulting differential cross section will look like figure 91. In reality of course the effective P pole must have an intercept $\alpha_P(0)$ just above 1 and a small but finite slope, and these are also reflected in the cut trajectory through (9.4); hence both contributions should have a modified energy dependence and a finite real part. The effect of this real part is to fill in the dip a little bit and ensure that the cross section does not vanish completely, which is what the data imply. The result is an entirely satisfactory account of the data as shown in figure 82 (Collins and Gault 1978, Donnachie and Landshoff 1984).

The main deficiency of this sort of model is that the relative magnitudes of A_1 and A_2, and a_1 and a_2, are not determined. Hence neither the position of the dip, nor the

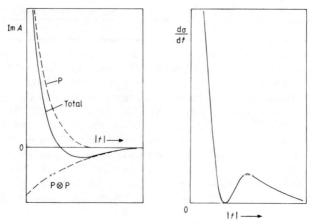

Figure 91. A schematic view of how the almost purely imaginary amplitudes for the Pomeron P and the P⊗P cut can interfere to produce the dip in the differential cross section.

exponential slope of the cross section beyond the dip ($|t| > 2$ GeV^2) where the double scattering amplitude dominates, can be predicted from the single-scattering P parameters (Collins 1977, ch 8). Similar dips are seen in the πp and Kp elastic scattering differential cross sections at $|t| \simeq 4$ GeV^2 which can easily be parametrised in a similar way, but again the position of the dip cannot be predicted *a priori*.

There have been many attempts to put these multiple-scattering ideas on a more formal basis—for example the geometrical model of Chou and Yang (1968, 1983), but without the addition of arbitrary parameters they predict a multiplicity of dips which are too close to the forward direction (Cheng *et al* 1973). There have also been proposals to interrelate the logarithmic changes in σ_T, σ(elastic), b etc through the concept of geometrical scaling (Dias de Deus and Kroll 1978), but so far this work has not achieved commanding plausibility through the accuracy of its predictions, and, perhaps more seriously, it is not at all clear how it is related to the fundamental features of hadronic structure. It is hard to see how any approach which ignores the Regge-like nature of small-$|t|$ scattering can be successful.

The behaviour of the large-$|t|$ differential cross section beyond the dip, though apparently exponential, is also quite compatible, within the errors in the data, with a power fall off of the form $d\sigma/dt \sim t^{-n}$, with $n \simeq 8$, at fixed s (Donnachie and Landshoff 1979). This behaviour was predicted by Landshoff (1974) as a consequence of the dominance of triple gluon exchanges as in figure 92(a). Unfortunately it is not known to what extent higher-order corrections like figure 92(b) will change this simple power-law behaviour through the appearance of so-called 'Sudakov form factors', which modify the point-like quark–gluon vertices upon which the prediction depends by logarithmic factors (Mueller 1981). Also, since gluons (like photons) have odd charge conjugation properties, and so couple with the opposite sign to quarks and antiquarks, this triple gluon exchange contribution will have the opposite sign in pp and p̄p scattering (Collins and Gault 1982). This would not affect the magnitude of the differential cross sections (which seem to be equal) in regions where the Landshoff mechanism is entirely dominant, but one would expect its interference with other contributions at smaller $|t|$ to result in some differences between high-energy pp and

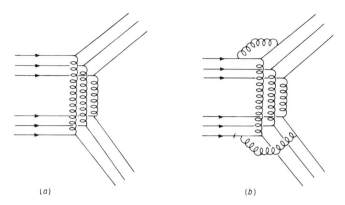

Figure 92. (a) The Landshoff triple-gluon exchange in pp scattering. (b) Vertex corrections (Sudakov form factors) which will modify the simple power behaviour of (a).

p̄p scattering and these have not been observed. There are, however, some recent results on p̄p scattering at large $|t|$ from the Collider (UA4 collaboration 1983) which suggests that the $|t| \simeq 1.4\,\text{GeV}^2$ dip is filled in at these very high energies. If so, this could be the first indication of a substantial change in the scattering mechanisms between high ($s \simeq 10^3\,\text{GeV}^2$) and super-high ($s \simeq 10^5\,\text{GeV}^2$) energies, and possibly of a difference between pp and p̄p of the type that the Landshoff mechanism anticipates. Time and more data will tell.

10.3   Particle production

Having looked at the total and elastic scattering cross sections we next want to examine particle production in soft hadron collisions.

Figure 93 shows how the average number of charged particles produced in an event increases with the collision energy. The new data from the CERN p̄p Collider indicate that $\langle n_{\text{ch}} \rangle$ is rising more rapidly than log s but certainly less rapidly than $s^{1/4}$. They are quite compatible in fact with $\langle n_{\text{ch}} \rangle \sim \log^2 s$, that is to say $\langle n_{\text{ch}} \rangle \sim \sigma_{\text{T}}$. In (2.9) we introduced the rapidity variable y. For the process $\text{AB} \to \text{CX}$ the rapidity of the produced particle C in the AB centre-of-mass frame is

$$y_{\text{C}} = \tfrac{1}{2} \log \left(\frac{E_{\text{C}} + p_{z\text{C}}}{E_{\text{C}} - p_{z\text{C}}} \right) = \tfrac{1}{2} \log \left(\frac{(E_{\text{C}} + p_{z\text{C}})^2}{m_{\text{TC}}^2} \right) \qquad (10.16)$$

where $m_{\text{TC}}^2 \equiv m_{\text{C}}^2 + p_{\text{T}}^2$. Now at very high energy the maximum and minimum values of y occur when $E_{\text{C}} \simeq \tfrac{1}{2}\sqrt{s}$, $p_{z\text{C}} \simeq \pm\tfrac{1}{2}\sqrt{s}$ and $p_{\text{TC}} \simeq 0$, so that the range of y within which C can be produced is

$$Y \equiv y_{\text{max}} - y_{\text{min}} \simeq \tfrac{1}{2} \log \frac{s}{m_{\text{C}}^2} - \tfrac{1}{2} \log \frac{m_{\text{C}}^2}{s} = \log \frac{s}{m_{\text{C}}^2} \qquad (10.17)$$

and so increases like log s. We showed in figure 13 that the y distribution of produced particles has a central plateau whose height increases slowly with s. Figure 94 indicates that the logarithmic rise of the central region continues up to Collider energies, and

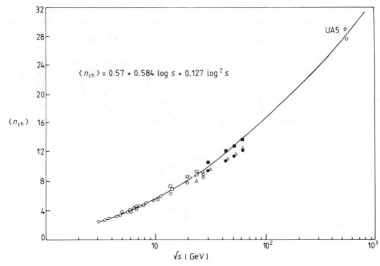

Figure 93. The average multiplicity of charged particles produced in pp scattering and a fit to the data showing that the leading term increases like $\log^2 s$, from Matthiae (1983).

so the $\sim\log^2 s$ increase of the multiplicity appears to be the combined result of the $\sim\log s$ increase of the width of the central plateau together with a $\log s$ increase of the plateau height. However, in the beam fragmentation regions $y \simeq y_{max}$ or y_{min} (i.e. $M^2/s \to 0$) there is little change in the multiplicity between ISR and Collider energies, and an approximate scaling of the quasi-elastic peak (see figure 95), so the fragmentation of the forward-going partons (see figure 15) appears to be independent of s, and it is in the central region, where the low-x sea partons interact, that the increase of multiplicity occurs. It is found that in very high multiplicity events the particle production is concentrated in the very low-y region, that is, most of the particles are produced nearly at rest in the centre-of-mass system.

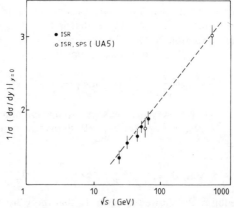

Figure 94. The height of the central plateau $(1/\sigma)\, d\sigma/dy$ at $y=0$ plotted as a function of $\sqrt{s}$ from the UA5 collaboration (1981).

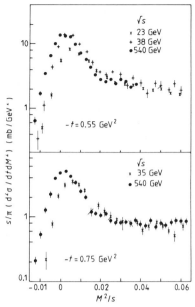

Figure 95. The invariant cross section for inelastic scattering $\bar{p}p \to \bar{p}X$ in the fragmentation region at the Collider compared with ISR data at lower energies showing (to within about 20%) approximate scaling. The figure is taken from Matthiae (1983).

Figure 96 demonstrates that the p_T distribution is broader at the Collider than at ISR showing that at higher energies there are more fairly hard parton collisions which produce higher p_T particles. In fact, it is the high multiplicity events which have the higher average p_T per particle. This effect was not observed at lower energies and could indicate that new production processes are occurring. However, these results are, at least qualitatively, just what one expects from QCD. If the forward-going partons are regarded as a jet, then the amount of gluon bremsstrahlung should increase with

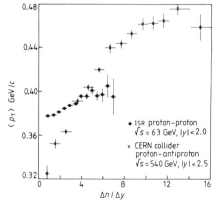

Figure 96. Growth of the average p_T with the multiplicity per unit rapidity interval at ISR and the Collider, from Breakstone *et al* (1983).

s, resulting in an increase of both $\langle n \rangle$ and $\langle p_T \rangle$ due to the formation of mini-jets with energies of a few GeV. It is found that if there is a high multiplicity in the forward direction then there is in the backward direction too, indicating that the harder interaction involves partons from both of the incoming particles. Similar high $\langle n \rangle$, high $\langle p_T \rangle$ events are also seen in the much higher energy cosmic ray collisions with air nuclei (Lattes *et al* 1980). Unfortunately, however, QCD only permits us to calculate the parton multiplicity and p_T distributions within isolated clusters of coherently interacting partons, and not the hadron multiplicity in correlated parton jets, so that a direct confrontation of these data with QCD is not really possible. It has been suggested that these high multiplicity events may be regarded as a consequence of the formation of a quark–gluon 'plasma' which radiates partons in the central region (Van Hove 1982). Alternatively they have been associated with the formation of further fragmenting strings of the type shown in figures 54(*a*) and 85(*c*) (Capella and Krzywicki 1983) but in this case the increased $\langle p_T \rangle$ should presumably occur outside the central rapidity region too. It may therefore be possible to test such models better when more detailed data become available.

At the ISR it is possible to compare pp and p̄p scattering and it is found that almost all aspects of the data, such as $\langle n \rangle$, $\langle p_T \rangle$ and the *x* distributions, are identical. The main difference, as one would expect, concerns the charges of the particles which are produced. Figure 97, which contrasts the *x* distributions of positively and negatively charged particles in pp and p̄p interactions, illustrates this clearly. In pp scattering (figure 97(*a*)) positive particles are produced in the central peak and in both fragmentation regions while negative particles occur mainly in the central region. But in p̄p

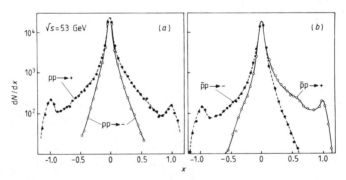

Figure 97. The *x* distribution of positive and negative particles produced in (*a*) pp and (*b*) p̄p collisions (Breakstone *et al* 1983).

collisions, although the positive particle distribution is the same as in pp for $x > 0$, for $x < 0$ it looks like that for negative particles. And negative particle production is a mirror image of that for positive, reflected about $x = 0$. This demonstrates conclusively that although in the fragmentation regions the particles produced are associated with the beams, in the central region they are completely independent of the beam particles, just as one would expect from figure 15. There is a short-range clustering of particles in rapidity but this is independent of the energy and seems to result from the decay of resonances.

It is found that there are significant differences between pp and p̄p scattering in the production of specific particles (Chauvat *et al* 1983), such as for example Δ^{++} or

Λ; Δ^{++} being suppressed and Λ enhanced in $\bar{p}p$ compared to pp particularly in the proton fragmentation region at large $|x|$. This is presumably because in $\bar{p}p$ scattering $\bar{q}q$ annihilation can occur leaving a forward-going diquark, which is more likely to be (ud) than (uu) because there are more $\bar{u}$s than $\bar{d}$s in a $\bar{p}$. And, as can be seen from table 2, (ud) can easily form a $\Lambda(uds)$ but not $\Delta^{++}(uuu)$. Also the yields of p and $\bar{p}$ are quite different even in the central region, indicating that perhaps a third of the central baryons are due to the stopping of one of the incoming baryons rather than the creation of a $\bar{p}p$ pair (Matthiae 1983).

Once these effects of the leading particle have been removed, however, the fragmentation of the hadronic system produced in hadron collisions seems remarkably similar to that found in deep inelastic scattering and $e^+e^- \to$ hadrons (Kowalski 1983, Basile *et al* 1983). So even if we cannot calculate the hadronisation process from first principles there can be little doubt that the same type of mechanism is occurring in all types of experiment.

The exchange of the Pomeron in figure 75 (the normal diagram) thus seems to be remarkably similar in its effect to the photon exchange of figure 25. This similarity of Pomeron and photon couplings has been exploited to predict the P coupling in various elastic and diffractive scattering processes with remarkable success (Collins and Wright 1979). It has been suggested by Donnachie and Landshoff (Landshoff 1983) that this analogy should also be exploited to relate hadronic fragmentation in hadron scattering to that in deep inelastic scattering.

In the region of very fast moving beam fragments where $x = (1 - M^2/s) \to 1$ (see (8.2)) the triple Regge diagram of figure 75 should dominate, leading to the prediction that the inclusive cross section behaves like (8.7), that is

$$f \sim (M^2)^{\alpha_P(0) - \alpha_i(t) - \alpha_j(t)} \sim (M^2)^{-1} \tag{10.18}$$

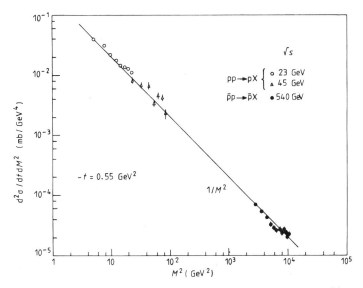

Figure 98. The cross section for pp $\to$ pX and $\bar{p}p \to \bar{p}X$ at various energies but fixed $|t|$ in the interval $0.01 \le M^2/s \le 0.04$ plotted against M^2, showing the $(M^2)^{-1}$ variation predicted by (10.18), from Matthiae (1983).

if the Pomeron with $\alpha_P(0) \simeq 1$ dominates all the exchanges. In figure 98 we show data for $pp \to pX$ and $\bar{p}p \to \bar{p}X$ with $0.01 \leqslant M^2/s \leqslant 0.04$ and see that the ISR and Collider results are in very good agreement with this prediction while the data range over several orders of magnitude.

Now if the P in figure 75 (normal) is replaced by a photon as in figure 25 we get

$$f \sim \frac{1}{s} \left(\frac{s}{M^2} \right)^{2\alpha_P(t)} f_p(x, q^2) \tag{10.19}$$

where f_p is the sum of the parton structure functions of the proton (cf (3.9)) with $t = q^2$ and $x = -t/M^2 \to 0$. As in (3.17) the x^{-1} behaviour of the sea structure function is related to the $(M^2)^1$ behaviour of the P in (10.18). With a more detailed knowledge of the structure functions for individual flavours of quarks it may be possible to make this relationship more quantitative.

Although much detailed information has yet to be obtained about the soft collision processes at these very high energies, and of course there may still be some surprises in store, it is already fairly clear that the main trends of such observables as σ_T, $d\sigma/dt$, $\langle n \rangle$, $\langle p_T \rangle$, and the x and y distributions continue up to collider energies the slow (logarithmic) evolution with energy that had been predicted on the basis of much lower energy experiments. An important achievement of the $\bar{p}p$ Collider, with its much larger value of log s, has been to establish more decisively the rate of this evolution.

11

JETS AT COLLIDER ENERGIES

The occurrence of jets of hadrons as a result of large-angle parton scattering is one of the most straightforward predictions of QCD, and so jet phenomena provide one of the principal ways in which QCD can be confronted with experimental data. At the p̄p collider very well defined wide-angle hadron jets are seen, and in this chapter we shall concentrate on how such data may be interpreted in terms of the underlying parton interactions. Good reviews of jet physics may be found in Giacomelli and Jacob (1979), Jacob and Landshoff (1978), Jacob (1983a, b), Wolf (1982), Söding (1983), Sosnowski (1983) and Rubbia (1983).

11.1 Calorimetric jet triggers

In chapter 2 we remarked how large-p_T hadrons are expected to be produced in four-jet events (see figure $17(b)$). Two of these jets result from hard, large-angle, parton scattering while the others contain the fragments of the beam particles. But in chapter 5 we demonstrated that, because of the fragmentation process, a high-p_T jet has only a small probability of producing a high-p_T particle. This is called the 'trigger bias' effect (see (5.16)). So if one identifies jets by looking for large-p_T particles one will inevitably miss most of the interesting jet cross section.

A more effective way of finding these high-p_T jets at high energy is to trigger instead on the occurrence of a large amount of energy (summed over all the particles) deposited in calorimeters at a fairly large angle to the beam direction. It is useful to introduce the transverse energy of particle i defined by

$$E_{iT} \equiv E_i \sin \theta_i \qquad (11.1)$$

where E_i is its energy and θ_i is the centre-of-mass angle between its direction of motion and the beam axis. Then the total transverse energy of an event is given by

$$E_T \equiv \sum_i E_{iT} \qquad (11.2)$$

summed over all the particles in the event.

Ideally, if all the spectator fragments go forward ($\theta = 0$) then E_T is the total energy of the scattered partons. In practice, for too low an E_T there will be a contamination by particles which are thrown off sideways by the spectator beams. Also since colour neutralisation requires that there be an interaction between the scattered partons and the spectator partons there will inevitably be some particles which cannot really be associated entirely with any of the four jets but which have some modest amount of transverse energy. And, of course, unless $E_T \gg \Lambda$, the hadronic mass scale, higher-order QCD effects and 'higher twists' necessarily are significant, but to an unpredictable extent. For all these reasons it is fairly obvious that to have well defined jets one must have $E_T \gg \Lambda$, m_{hadron} and $\langle p_\perp \rangle$ to the jet axis, but not at all clear a priori just how large E_T needs to be.

In experiments at the CERN SPS and FNAL covering the range $100 < p_{lab} < 400\ GeV/c$, and in early ISR work on pp scattering ($20 < \sqrt{s} < 60\ GeV$), although there were certainly plenty of high-E_T events, rather few of them looked as though they had the expected four-jet structure.

One way of quantifying the 'jettyness' of the data is to note that in an ideal four-jet event all the incoming and outgoing partons will lie in a plane, and so, neglecting any momentum the hadrons may have transverse to their jet axis, all the particles produced with large E_T should lie in a single plane, which will also contain the beam axis (see figure 99). One can define the degree of planarity, P, by

$$P \equiv \frac{A - B}{A + B} \tag{11.3}$$

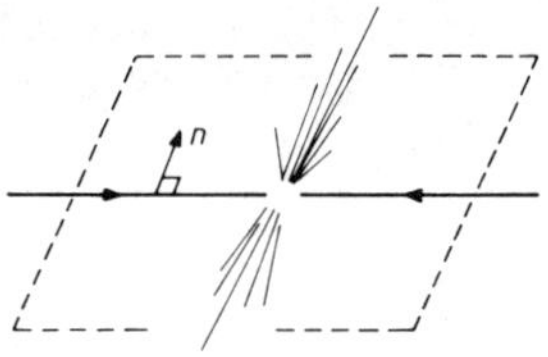

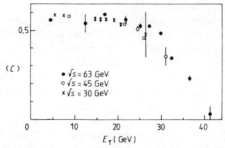

Figure 99. A scattering process producing two large p_T jets. n is a unit vector perpendicular to the beam axis which lies in the plane containing both the beams and the jets.

where A is given by the sum of the momentum components of all the particles in this plane

$$A \equiv \max \left(\sum_i (\boldsymbol{p}_{iT} \cdot \boldsymbol{n})^2 \right) \tag{11.4}$$

while B is the sum of the components perpendicular to this plane

$$B = \min \left(\sum_i (\boldsymbol{p}_{iT} \times \boldsymbol{n})^2 \right). \tag{11.5}$$

The direction of the unit vector $\boldsymbol{n}$, which is perpendicular to the beam axis, is chosen to maximise A and minimise B. Clearly for perfect jets there would be a direction $\boldsymbol{n}$ such that $B = 0$ giving $P = 1$, whereas if the particles were produced isotropically about the beam axis we would find $P = 0$ for any direction of $\boldsymbol{n}$.

Figure 100 shows the circularity $C (C \equiv 1 - P)$, which is the two-dimensional equivalent of sphericity (5.1), at various ISR energies. Evidently C remains large,

Figure 100. The average circularity $\langle C \rangle$ versus E_T at three ISR energies. Only for $E_T > 30\ GeV$ does the circularity become small indicating the dominance of two high-p_T jets (AFS collaboration 1983a).

indicating a fairly isotropic distribution of particles, until $E_T \simeq 30 \, \text{GeV}$ when, remarkably, it suddenly starts to fall quite dramatically. It would seem that events containing a high multiplicity of particles, each with modest E_{iT}, dominate the event rate until this very high total E_T is achieved. It is not surprising therefore that in experiments with $\sqrt{s} < 30 \, \text{GeV}$ it is difficult to extract a clean four-jet signal from this 'noise', but quite remarkable how once $E_T > 30 \, \text{GeV}$ really clean jets emerge.

At the $\bar{p}p$ Collider, however, there is plenty of energy available to produce transverse jets which stand out very clearly from the background (UA2, UA1 collaborations 1982, 1983b). Two examples are shown in figure 101 where the angular clustering

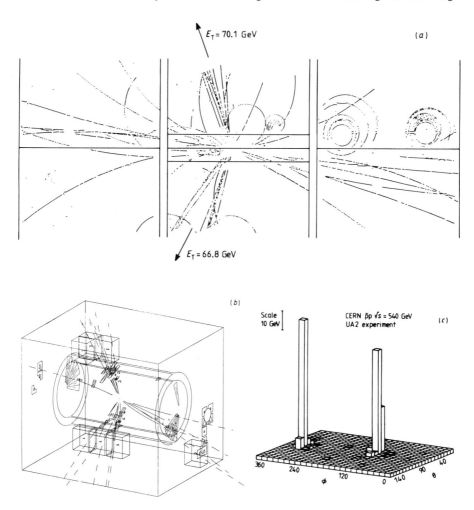

Figure 101. Jet events observed by the UA1 and UA2 collaborations at CERN. Diagram (a) shows the particle tracks as reconstructed in the UA1 central detector, and (b) shows the same event with the particle hits in the outer calorimeters also indicated, but leaving out tracks of low transverse momentum. Plot (c) corresponds to a two-jet event observed by the UA2 collaboration, and shows the transverse energy deposited in the UA2 detector as a function of the polar angle θ and azimuth ϕ.

of the particles into two wide-angle jets is quite unambiguous. In fact for sufficiently large E_T almost the whole of the transverse energy is carried by particles from these two scattered jets, which share the energy equally, as figure 102 illustrates. The experiments performed at the Collider have thus put the occurrence of wide-angle jets in hadron scattering beyond dispute.

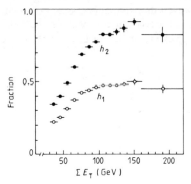

Figure 102. The fraction of the total E_T carried by the jet with the highest E_T and by the two jets with the highest E_T, as a function of the total E_T. The fractions are denoted h_1 and h_2 respectively. It will be seen that for $E_T > 150$ GeV almost all the transverse energy is contained within two jets (UA2 collaboration 1983a).

11.2 The jet cross section

From (3.4) the cross section for producing a parton jet of momentum $\boldsymbol{p}_c$ and energy E_c in the process $AB \to (\text{jet})_c X$ is given by

$$E_c \frac{\mathrm{d}^3\sigma}{\mathrm{d}^3 p_c} = \sum_{ab} \int_0^1 \mathrm{d}x_a \int_0^1 \mathrm{d}x_b\, f_A^a(x_a) f_B^b(x_b) \frac{1}{\pi} \frac{\mathrm{d}\sigma}{\mathrm{d}\hat{t}} (ab \to cd) \tag{11.6}$$

with the various parton-scattering cross sections given in table 3. In the centre-of-mass system the two incoming particles, A and B in figure 24, have four-momenta $(\frac{1}{2}\sqrt{s}, 0, \frac{1}{2}\sqrt{s})$ and $(\frac{1}{2}\sqrt{s}, 0, -\frac{1}{2}\sqrt{s})$ respectively and so the partons a and b have the four-momenta $x_a(\frac{1}{2}\sqrt{s}, 0, \frac{1}{2}\sqrt{s})$ and $x_b(\frac{1}{2}\sqrt{s}, 0, -\frac{1}{2}\sqrt{s})$ respectively. For simplicity we shall consider the case where the partons are scattered at right angles to the beam direction, which requires that $x_a = x_b \equiv x$ and so the energy, E_T, and momentum magnitude, p_T, of each of the scattered partons, c and d, is $E_T = p_T = x\sqrt{s}/2$ (see figure 103). Remember that for all these kinematical calculations we neglect the masses of

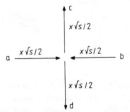

Figure 103. Parton scattering $ab \to cd$ at 90°. Neglecting the masses, each parton carries a fraction x of the momentum of the incoming particles, A and B, which each have energy $\sqrt{s}/2$.

the particles and the partons, which are very small compared to their momenta. It is usual to introduce

$$x_T \equiv \frac{2p_T}{\sqrt{s}} \tag{11.7}$$

so that for $90°$ scattering $x_T = x$. Then from (3.3) we find that the subprocess variables are

$$\bar{s} \simeq x^2 s \qquad \bar{t} \simeq -\tfrac{1}{2}x^2 s = -2p_T^2. \tag{11.8}$$

At the ISR the maximum energy is $\sqrt{s} = 62$ GeV and so if we require say $E_T \simeq 2p_T > 30$ GeV to distinguish the jets clearly with very little background (i.e. small circularity C), then we must have $x > 0.5$. But figure 27 indicates that there are relatively few partons with such a large x, just a few valence quarks, and so the jet cross section is small. But at the Collider, where $\sqrt{s} = 540$ GeV, we only need $x > 0.1$, and in the small-x region there is a very large number of sea partons, predominantly gluons, and so the jet cross section is expected to be very much larger.

Figure 104 is a comparison of data taken at the ISR and the Collider with the cross sections predicted from (11.6). Of course the gluon structure functions are known only rather indirectly from scaling violations in deep inelastic scattering experiments, but despite this the agreement is excellent. Because of the lower x values required, the Collider cross sections are some 10^3 times as big, for the same E_T, as those found at ISR. Also the jet cross section varies by a factor of about 10^6 over the E_T range measured at the Collider, yet nowhere do the crude QCD predictions differ from the data by more than a factor of 2–3. This is a remarkable success for the parton approach to jet cross sections.

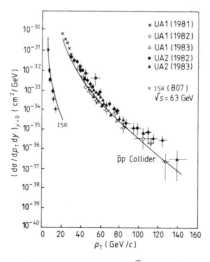

Figure 104. Comparison of the jet cross sections at ISR ($\sqrt{s} = 63$ GeV) and the Collider ($\sqrt{s} = 540$ GeV) with standard QCD predictions, taken from Jacob (1983b).

The scaling of the parton cross sections in table 3 leads one to expect that

$$\frac{d\sigma}{dp_T^2} = \frac{1}{p_T^4} f[x_T, \alpha_s(p_T^2)] \tag{11.9}$$

and so at fixed x_T the cross section should fall with p_T like p_T^{-4}, but with an additional logarithmic decrease due to the variation of $\alpha_s(p_T^2)$. However, since the product of the two structure functions in (11.6) is a rapidly varying function of x_T, the p_T dependence at fixed s falls much faster, approximately like p_T^{-9}. At the lower values of E_T the dominant contribution is from the parton scattering process $gg \rightarrow gg$, with $gq \rightarrow gq$ and $g\bar{q} \rightarrow g\bar{q}$ taking over at intermediate values and then for $x_T > 0.3$ valence quark–antiquark scattering, $q\bar{q} \rightarrow q\bar{q}$, becomes dominant, as a sample calculation from Antoniou et al (1983) in figure 105 shows.

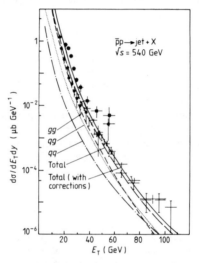

Figure 105. A QCD calculation of the cross section for $\bar{p}p \rightarrow jet + X$ at the Collider showing the contributions of the various subprocesses $q\bar{q} \rightarrow q\bar{q}$, $qg \rightarrow qg$ and $gg \rightarrow gg$, from Antoniou et al (1983).

The process $q\bar{q} \rightarrow q\bar{q}$ in table 3 behaves like

$$\bar{t}^{-2} \sim (1 - \cos\theta)^{-2} \sim \left(\sin^4\frac{\theta}{2}\right)^{-1} \tag{11.10}$$

which is the angular distribution predicted by Rutherford for the scattering of classical point-like charged particles. The other important processes, $qg \rightarrow qg$, $\bar{q}g \rightarrow \bar{q}g$ and $gg \rightarrow gg$, have a very similar structure and so one may expect approximately this angle-dependence for all the jets produced at the Collider (Halzen and Hoyer 1983, Combridge and Maxwell 1984). This expectation is well verified in figure 106 which demonstrates very clearly that the jets are the result of the scattering of point-like partons mediated by vector gluon exchange. In fact these experiments have been used

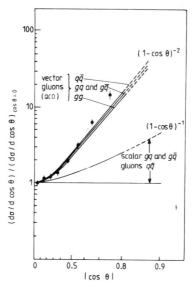

Figure 106. The angular distribution of 2-jet events (UA1) compared to the predictions of parton scattering which all give approximately the $(\sin^4 \theta/2)^{-1}$ distribution of Rutherford scattering (Rubbia 1983).

to confirm that not only are gluons definitely spin 1 particles, but that quarks are point-like down to a distance of about 5×10^{-19} m.

Another important consequence of the similar structure of the parton subprocesses is that the cross section (11.6) is predicted to factorise into the product of contributions from each of the two hadrons, i.e. $f_A^a(x_a) f_B^b(x_b)$. In fact, because of the relative magnitude of the colour factors in table 3, to a good approximation over a large part of the angular range, including 90° scattering, the subprocesses contribute in the ratio

$$gg \to gg : qg \to qg : \bar{q}g \to \bar{q}g : q\bar{q} \to q\bar{q} = 1 : \tfrac{4}{9} : \tfrac{4}{9} : (\tfrac{4}{9})^2 \qquad (11.11)$$

so effectively one sees a weighted mixture of the quark and gluon densities in each hadron, of the form

$$f(x) = f_p^g(x) + \tfrac{4}{9} f_p^q(x) \qquad (11.12)$$

so that if A is the proton and B the antiproton then

$$\frac{d\sigma}{dx_a \, dx_b} \sim (f_p^g(x_a) + \tfrac{4}{9} f_p^q(x_a))(f_p^g(x_b) + \tfrac{4}{9} f_p^{\bar{q}}(x_b)). \qquad (11.13)$$

One consequence of this is that it will be very difficult to test the admixture of these QCD subprocesses unless one can distinguish quark jets from gluon jets. Figure 107 shows that the cross section does indeed factorise and that the shape of the weighted structure function combination is similar to those deduced from deep inelastic lepton scattering experiments.

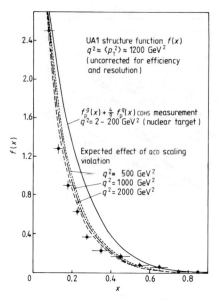

Figure 107. The structure function combination $f(x) = f_p^g(x) + \frac{4}{9} f_p^q(x)$ as observed by the UA1 collaboration (Rubbia 1983) compared with deep inelastic scattering results, including the effect of scaling violations, with various assumptions for the effective value of $q^2 \equiv Q^2$, see (4.4) and (4.5).

A significant fraction of the events at the collider show three jets outside the beam fragmentation regions (see figure 108). In fact, in about 10% of the cases in which the two dominant jets each have $E_T(\text{jet}) > 15 \text{ GeV}$ there is a third jet with $E_T > 15 \text{ GeV}$ too (UA1 collaboration 1983c). These additional jets are presumably the result of gluon bremsstrahlung from one of the scattered quarks or gluons (like figure 43) and which we know from e^+e^- experiments should occur at a rate of order α_s with respect to that for producing two wide-angle jets (see (5.5)). This is confirmed by the

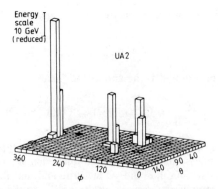

Figure 108. An event with three high-p_T jets observed by the UA2 collaboration showing the transverse energy deposition in the detector as a function of the polar angle θ and azimuth ϕ. Compare this with figure 101(c).

observation that the angular position of the third jet is strongly concentrated in the vicinity of the less energetic of the two leading jets as one would anticipate from figure 43.

It has thus been verified that all the main expectations of QCD as regards the cross sections for producing jets in hadron scattering are fulfilled at Collider energies. Much more detailed studies are still needed to demonstrate that QCD is uniquely the correct explanation, however.

Eventually it should be possible to find events in which there are four wide-angle jets due to double scattering as illustrated in figure 109. Such processes cannot dominate at very large E_T but should provide a significant number of above-average multiplicity events at moderate E_T. Already there is a little evidence for such multiple-scattering processes (Jacob 1983b), and they could possibly be related to the very high multiplicities which have been observed in some cosmic ray collisions (Lattes *et al* 1980).

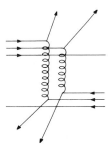

Figure 109. Double parton scattering producing four high-p_T jets.

11.3 Jet fragmentation

It is found that the hadrons in a Collider jet have only a rather small momentum component perpendicular to the jet axis, the average value being $\langle p_T \rangle \simeq 0.6 \, \text{GeV}/c$ (UA1, UA2 collaborations 1983a, 1983a), and hence the jets occupy only a very small angular region, as in figures 101 and 108. In fact the p_T distribution is remarkably similar to that found at ISR and in $e^+e^- \to q\bar{q}$ even though we expect a predominance of gluon jets, not quark jets, at the Collider. There should be a slow increase of $\langle p_T \rangle$ with the energy of the jet as in e^+e^- scattering (see figure 45 and (5.8)), but these effects are only just beginning to show up in hadron scattering.

The fragmentation of the jet into charged particles is described by $D^{\text{ch}}(z)$ where

$$z \equiv \frac{p_L^{\text{ch}}}{E_{\text{jet}}} \tag{11.14}$$

is the fraction of the total energy of the jet (E_{jet}) which is taken by the longitudinal momentum component (along the jet axis, p_L) of the particular charged particle which is detected (see the discussion below (3.1)). This fragmentation function is given by

$$D^{\text{ch}}(z) = \frac{1}{N_{\text{jets}}} \frac{dn_{\text{ch}}}{dz} \tag{11.15}$$

where N_{jets} is the number of jets and dn_{ch} is the number of charged particles detected in the interval dz. Figure 110 shows that this fragmentation function for the predominantly gluon jets at the collider is identical (within admittedly rather large errors) to that found in the predominantly quark jets in high energy $e^+e^- \to$ hadrons. There is rather little evidence as yet for scaling violations, i.e. $D(z) \to D(z, E_{\text{jet}})$ as in (4.7) and (4.8), but such effects have been detected in e^+e^- jets (Wolf 1982).

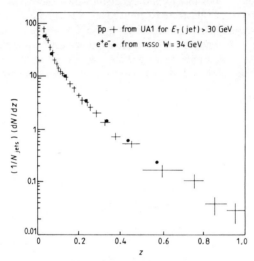

Figure 110. The fragmentation function of the gluon jets at the Collider obtained from (11.15) by the UA1 collaboration (1983a) compared with the fragmentation of quark jets in $e^+e^- \to$ hadrons.

The average multiplicity of particles produced at the collider appears to continue the same rising trend with energy as is found in e^+e^- collisions, as figure 111 demonstrates. The leading log approximation to the multiplicity of partons is given by QCD (Bassetto *et al* 1980, Furmanski *et al* 1979) as

$$\langle n \rangle = n_0 + a \exp\left(b\sqrt{\log\left(s/\Lambda^2\right)}\right) \tag{11.16}$$

where n_0, a and b are parameters, $\sqrt{s}$ is the total energy of the jet and Λ is the QCD energy scale. If the production of hadrons reflects that of partons then one may anticipate a similar energy dependence of the hadron multiplicity. The data are reasonably consistent with this view, but the errors are necessarily rather large, not least because of the model dependence involved in deciding which particles should be included in the jet.

In part this model dependence arises because in order to decide which hadrons are to be associated with a given jet one is obliged to make assumptions about the way in which the hadrons are actually produced as the jet develops. Two simple viewpoints which have been employed in many analyses of jet data are (i) the independent fragmentation model of Feynman and Field (1978) (see also Hoyer *et al* 1979), which is embodied in formulae like (3.4) and (3.19), in which it is supposed that each parton

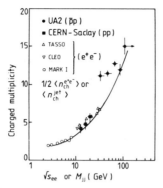

Figure 111. The average multiplicity of charged particles in a jet as a function of the energy of the jet compared with that found in e^+e^- annihilation, and compared with the theoretical prediction of (11.16). The figure is from the UA2 collaboration (1983a).

fragments separately into more partons and ultimately into hadrons (see figure 112(a)); and (ii) the colour string model of Andersson *et al* (1980, 1983) which was described in § 1.4, in which the partons are regarded as the ends of strings of the colour field, and, on stretching, these strings break up by creating $q\bar{q}$ pairs out of the vacuum (figure 112(b)).

For a two-jet event like $e^+e^- \to q\bar{q}$ there is little difference between these pictures because the string lies along the direction of motion of the quarks, but in multi-jet events like figure 112 there is a difference in that in (a) the hadrons are produced entirely along the line of motion of the partons while in (b) they are produced along a line joining two of the partons, presumably approximately a straight line in the rest frame of those two partons.

(a) (b)

Figure 112. A 3-jet event according to (a) the independent fragmentation model, and (b) the colour string model.

Unfortunately these differences mainly relate to the production of the slower hadrons whose association with jets is in any case ambiguous. By making a suitable choice of the value of α_s both models seem able to fit all the data on $e^+e^- \to$ jets (Söding 1983). However, ISR data on the correlation between the trigger particle, C, and a secondary particle, D, in the process pp→CDX appear to favour a picture in which the colour string connects the hard trigger-particle parton with the spectators in the opposite hadron, as in figures 54 and 85(c) (Wegener 1983).

Further information about the hadronisation mechanism can be obtained from more detailed studies of the fragmentation into different types of particle, which we discuss next.

11.4 Particle production in jets

At the ISR we expect that most of the jets will result from the scattering of the valence quarks. Since the valence quark structure of a proton is uud, a π^+ ($u\bar{d}$) can be made from a scattered u quark which picks up a $\bar{d}$ from the vacuum (figure 113) while a

Figure 113. A u quark fragmenting into a π^+ meson.

π^- ($d\bar{u}$) can arise from a scattered d quark. Hence one might expect that a π^+ is twice as likely to be found as a π^- among the very high-p_T particles which are the leading particles in the jets. As figure 114 shows, the cross section ratio $d\sigma(\pi^+)/d\sigma(\pi^-)$ is close to 1 at small x_T where production of both q and $\bar{q}$ out of the vacuum is dominant, but tends to 2 at large x_T where the charge of the leading quark is more important.

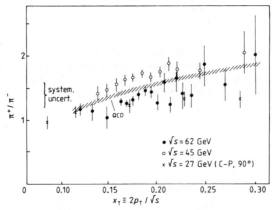

Figure 114. The cross section ratio $d\sigma(\pi^+)/d\sigma(\pi^-)$ at the ISR as a function of x_T (SFM collaboration 1984). At small x_T these cross sections are equal but the ratio tends to 2, as expected for valence quark scattering, as x_T increases.

Alternatively the leading u quark could pick up an $\bar{s}$ from the vacuum and produce a K^+, but this process is somewhat suppressed because the strange quark is heavier than a u or d. Figure 115 shows that the ratio of K^+/π^+ is about 0.45, independent of x_T, in these experiments. This seems to differ somewhat from the suppression factor of 0.3 found in central production, as discussed in chapter 6. On the other hand K^- ($\bar{u}s$) contains none of the valence quarks of the proton so its production should be highly suppressed at large x_T, as one finds in figure 115.

The production of a positively charged particle, as in figure 113, leaves behind a d quark which is more likely to produce a negatively charged particle. Hence if figure 113 is not too misleading a view of how the particles are manufactured, one would expect to find pairs of particles with similar momenta but opposite charges. This sort of local 'compensation' of charge by particles which have very similar rapidities (along

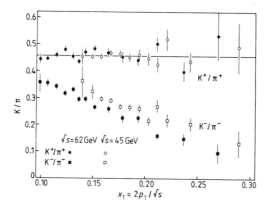

Figure 115. Cross section ratios measured at the ISR as a function of x_T for $d\sigma(K^+)/d\sigma(\pi^+)$ and $d\sigma(K^-)/d\sigma(\pi^-)$, SFM collaboration (1984). The approximate equality of the ratios at small x reflects $s\bar{s}$ production out of the vacuum.

the jet axis) has been well verified. But of course much of the observed clustering of particles is due to resonance decay, such as $u \to \rho^+ \to \pi^+ \pi^0$ rather than $u \to \pi^+$ directly, which greatly confuses any attempt to establish correlations between the parton charges.

To produce a proton we need either a diquark jet, which seems likely only if the diquark system has been scattered at a very small angle because of the effect of the diquark's form factor (see (2.22)), or the quark jet must pick up two quarks (see figure 116). On the other hand to produce a $\bar{p}$ from proton scattering demands that all the valence antiquarks must be created in the vacuum. Figure 117 demonstrates that while the proportion of protons in the jet decreases with increasing jet angle the proportion of antiprotons is more or less constant, and similar to the proportion of protons in large-angle jets. This suggests that wide-angle protons are produced predominantly either from gluons which turn into $p\bar{p}$ pairs, as in figure 116(c), or that the creation of a proton from a quark jet (figure 116(b)) leaves behind an anti-diquark (as opposed to two independent antiquarks) which has a high probability of fragmenting into a $\bar{p}$.

Much more detailed studies of the particle content of quark jets have been made in $e^+e^- \to q\bar{q}$ annihilation experiments (Wolf 1982, Söding 1983). Figure 118 illustrates that while the low-momentum hadrons are mainly pions, presumably because unenergetic quarks can only create very light hadrons, at the highest momenta the pion and kaon production rates have become more nearly equal, with at least a 10% rate of

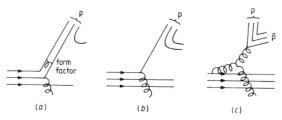

Figure 116. The production of a proton from (a) a diquark jet, (b) a quark jet and (c) a gluon jet. In (a) we show the hard gluon contribution to the form factor.

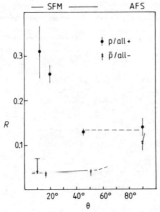

Figure 117. The ratio of the proton and antiproton production cross sections to the total inclusive cross section as a function of the scattering angle for pp collisions at $\sqrt{s} = 63$ GeV, taken from Sosnowski (1983).

producing baryons. At a total centre-of-mass energy of 34 GeV it is found that on average about 21 particles are produced, of which 13 are charged. But a very large fraction, perhaps as much as 90%, of the pions seem to come from the decays of heavier states, ρ's, ω's and K*'s for example, so there are probably only about 10 primary hadrons produced, i.e. an average of 9 $q\bar{q}$ pairs produced in addition to the initial $q\bar{q}$ from the decay of the virtual photon. The K/π ratio is similar to that found in the jets produced by scattering hadrons. The occurrence of 10% of baryons seems to imply that there is about one chance in ten that a diquark–anti-diquark (qq–$\bar{q}\bar{q}$) is

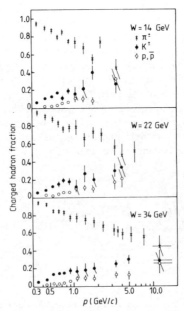

Figure 118. The fraction of charged pions, kaons and p, $\bar{p}$ as a function of momentum in $e^+e^- \to$ hadrons, from the TASSO collaboration (1983).

produced in the vacuum in preference to a $\bar{q}q$ pair. About 30% of the baryons are strange, mainly Λ's with a few Ξ's. There is very good evidence for the short-range (in rapidity) compensation of charge, strangeness and baryon number. Thus, for example, if a proton is produced there is nearly always an antiproton in the same jet with a very similar rapidity along the jet axis. This is just what one expects from figure 116(b). If alternatively one had diquark production at the photon vertex, i.e. $e^+e^- \rightarrow \gamma \rightarrow$ $(qq)(\bar{q}\bar{q})$, one would expect the baryon and antibaryon to appear in opposite jets. Presumably the form factor at the photon–diquark vertex suppresses this sort of contribution at high energy.

In very high energy e^+e^- collisions the heavier c and b quarks are plentifully produced and it is possible to observe their fragmentation into heavy particles; $c \rightarrow D$ and $b \rightarrow B$. Because the quark mass is not much less than that of the hadron we expect that this leading hadron will carry a large fraction of the quark's momentum; unlike the pions which dominate the fragmentation of the light u and d quarks (see (3.25)). It is found that on average the D meson from a c quark acquires nearly 70% of its momentum while the average B has about 80% of the momentum of the b quark (Söding 1983). When it is remembered that many of the observed particles probably result from decays of excited states, for example $c \rightarrow D^* \rightarrow D\pi$, and the other decay products are also taking part of the quark's momentum, it is clear that the initial fragmentation, $c \rightarrow D^*$, must have produced very fast D^*'s.

There are also three-jet events resulting from the process $e^+e^- \rightarrow q\bar{q}g$ as in figure 43. These should enable us to discover how gluon jets differ from quark jets, provided we can distinguish which jet belongs to which parton. Since the gluon is radiated from one of the quarks it will usually provide the jet with the lowest energy. Another way to observe gluon jets is to detect the hadrons from ψ (or Υ) decays which we expect to be predominantly $\psi \rightarrow ggg \rightarrow$ hadrons since the usual type of vector meson decay $\psi \rightarrow c\bar{c} \rightarrow D\bar{D}$ is impossible because $m_\psi < 2m_D$. Unfortunately the ψ and Υ masses are too low for three clean jets to be resolved but if bound states of the much heavier $t\bar{t}$ system can be produced (remember $m_t > 20 \text{ GeV}/c^2$) we would expect to see three energetic jets from their decay. Also we have argued in § 11.2 that at the Collider most of the jets must be gluon jets, rather than quark jets, because they result from partons with small $|x|$, so a study of the differences between ISR and Collider jets should be instructive.

The colour factors at the vertices give the quark–gluon coupling as $\frac{4}{3}\alpha_s$ (see (1.9)) but the triple–gluon coupling is $3\alpha_s$. Hence a gluon jet should be able to radiate $\frac{9}{4}$ times as many soft gluons as can a quark jet. If this is reflected in the hadron multiplicity we can expect that a gluon jet may contain $\frac{9}{4}$ times as many hadrons and that its transverse momentum distribution will be correspondingly wider. There is very little evidence for such an enhancement of multiplicity when one compares the gluon jets at the Collider with the quark jets of the ISR or with those from e^+e^- experiments in figure 110. However, it appears that the least energetic jet in the $e^+e^- \rightarrow$ three-jet events does have a somewhat greater spread of p_T ($\langle p_T \rangle_g \simeq 1.2\langle p_T \rangle_q$) and these gluon jets also seem to be more effective in producing baryons ($p\bar{p}$ or $\Lambda\bar{\Lambda}$) than are quark jets.

We conclude that within the accuracy of the available data the way in which particles are produced in jets is remarkably consistent with intuitive pictures like figure 116 as regards quantum number effects, and that the jets seen in such very different processes as hadron scattering (figure 17(a)), deep inelastic scattering (figure 20) and $e^+e^- \rightarrow$ hadrons (figure 41) are quite clearly manifestations of the same underlying mechanism: the hadronisation of fast-moving but confined partons. The parton model dressed by QCD seems indispensable to the analysis of all these types of process.

11.5   Direct photon production

The ambiguities which arise when one tries to use either large-p_T hadron production events or hadronic jets to try to probe the parton scattering processes which underlie hadron collisions stem largely from our uncertainty as to how the fragmentation of quarks or gluons into hadrons actually proceeds. The fragmentation functions $D(z, Q^2)$ are merely parametrisations of this process and their accuracy and range of applicability are hard to assess. It is very useful, therefore, that there is a way of circumventing all these problems by observing instead photons which emerge directly from the basic parton hard scattering processes (see Halzen and Scott 1978).

The lowest-order parton reaction which will produce a photon in pp scattering is the so called 'Compton scattering' process $gq \rightarrow \gamma q$ shown in figure 119. Apart

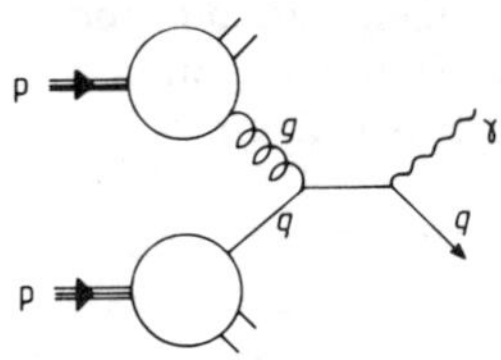

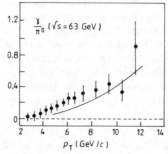

Figure 119. The 'Compton scattering' process $gq \rightarrow \gamma q$ for direct photon production in pp collisions.

from the replacement of the gluon–quark coupling $\frac{4}{3}\alpha_s$ by the photon–quark coupling $e_q^2\alpha$, the rate for this process should be essentially the same as that for $gq \rightarrow gq$ which is one of the dominant processes producing jets at the Collider. The photon production rate should thus be about 10^{-2} of the rate for producing gluon jets. We have found that the trigger bias effect (5.16) reduces the yield of single particles at high p_T by a factor of the order of 10^{-2} to 10^{-3} relative to the jet cross section because no single particle is likely to carry a sizable fraction of the jet's momentum. Hence one expects that at high p_T the ratio of the photon cross section to the single-pion cross section, for example, will be of order 1, not of order α as one finds in soft processes.

Of course there are many other sources of photons, such as the decays of hadrons like π^0 or $\eta \rightarrow \gamma\gamma$, $\omega \rightarrow \pi\gamma$, $\eta' \rightarrow \rho^0\gamma$, or the bremsstrahlung of photons by any of the charged particles which take part in the scattering process, but these mechanisms will

Figure 120. The ratio of single photons to π^0 mesons produced at the ISR as a function of p_T (AFS collaboration 1983b) compared with the QCD calculation of Benary et al (1983).

only produce photons copiously with small p_T, and so at very large p_T the Compton process and other QCD subprocesses, such as $qq \rightarrow qq\gamma$ and $q\bar{q} \rightarrow \gamma g$, should emerge as the dominant ones. In figure 120 we show some ISR data for the ratio of single photons to pions produced at 90° in proton–proton collisions, as a function of p_T, and it will be seen that for $p_T > 4$ GeV/c the Compton cross section, calculated from figure 119 using the known quark and gluon structure functions of the proton, is in good agreement with the data. The ratio of photons to π^0 mesons becomes of order 1 for $p_T \simeq 12$ GeV/c. At the Collider the prompt photon signal should be even clearer, and should among other things provide us with a very nice way of isolating the various QCD subprocesses.

12

W AND Z BOSONS

Radioactive decays such as $n \to pe^- \bar{\nu}$ depend on the weak nuclear force, which is quite different from the strong (colour) nuclear force which binds quarks to form hadrons. The development of the theory of this weak interaction is one of the most exciting and interesting stories in physics, culminating in 1983 with the discovery of the $W^\pm$ and Z^0 weak bosons at the CERN $\bar{p}p$ Collider with masses

$$M(W^\pm) = (81 \pm 2) \text{ GeV}$$
$$M(Z^0) = (93 \pm 2) \text{ GeV}$$

(12.1)

which, as we shall see, are exactly as predicted by the standard electroweak gauge theory.

The story starts in 1932 with Fermi's proposal of the four-fermion form of the weak interaction (to explain $n \to pe^- \bar{\nu}_e$). It then took 25 years of detailed experiments to show that, as far as low-energy weak interactions were concerned, Fermi had been almost correct. The only aspect he had not foreseen was parity violation: that only left-handed fermions (and right-handed antifermions) appear to participate in (charge-changing) weak interactions. But, despite its successful description of weak interaction phenomena, his theory was unsatisfactory for two closely related reasons. First, at sufficiently high energies (~ 300 GeV) the theory violates unitarity in that the $l = 0$ partial wave amplitude becomes > 1, and second, any attempt to calculate other than the lowest-order diagrams is plagued by infinities which cannot be removed by renormalisation. Unlike QED, the theory is not renormalisable. A modification of the theory is needed to overcome these problems and yet maintain the successful low-energy predictions. Gauge theory provided the answer. A complete discussion of weak interactions is outside the scope of this book, but in the next section we give a brief survey of the main ideas. For a more detailed account see Okun (1982) Bilenky and Hosek (1982), Bailin (1982), Aitchison and Hey (1982) or Halzen and Martin (1984).

12.1 Résumé of gauge theories and electroweak interactions

Nowadays it is widely believed that the structure of all the particle interactions can be determined by imposing local gauge symmetries. If the basic free-particle Lagrangian is required to be invariant under local gauge transformations then we must introduce new (gauge) fields interacting with the original fields in a specified way. A gauge transformation is a historical misnomer for a phase transformation of the form

$$\psi \to e^{i\phi(x)}\psi$$

(12.2)

and 'local' means that the arbitrary phase ϕ can depend on the particular space–time point. The set of all such transformations (12.2), which depend on a single function, form a U(1) gauge symmetry group.

QED is a U(1) gauge theory of this type. To see that this is so we start with the Lagrangian for a free electron of mass m and wavefunction $\psi(x)$ which obeys the Dirac equation, that is

$$\mathcal{L} = i\bar{\psi}\gamma^\mu \partial_\mu \psi - m\bar{\psi}\psi \qquad (12.3)$$

where $\partial_\mu \equiv \partial/\partial x^\mu$. Although $\mathcal{L}$ is invariant under 'global' phase transformations in which ϕ is independent of x, it is not invariant under (12.2) on account of the term containing $\partial_\mu \phi$. However local gauge invariance can be achieved by replacing ∂_μ in (12.3) by the so-called 'covariant' derivative

$$D_\mu \equiv \partial_\mu - ieA_\mu \qquad (12.4)$$

provided that the newly introduced field, the gauge field A_μ, transforms as

$$A_\mu \rightarrow A_\mu + \frac{1}{e}\partial_\mu \phi.$$

The second term precisely cancels the unwanted $\partial_\mu \phi$ term in the original Lagrangian and ensures the gauge invariance of (12.3). We see that the transformation property of A_μ is just the usual gauge freedom of the electromagnetic potential. In this way we are led to the Lagrangian of QED

$$\mathcal{L} = \bar{\psi}(i\gamma^\mu \partial_\mu - m)\psi + e\bar{\psi}\gamma^\mu A_\mu \psi - \tfrac{1}{4}F_{\mu\nu}F^{\mu\nu} \qquad (12.5)$$

where the last term is the Lorentz and gauge invariant construction of the kinetic energy of the A_μ field, involving the gauge invariant field strength tensor

$$F_{\mu\nu} \equiv \partial_\mu A_\nu - \partial_\nu A_\mu.$$

The gauge field, A_μ, may thus be identified with the photon field. Note that the addition of a mass term, $\tfrac{1}{2}M^2 A_\mu A^\mu$, to (12.5) is incompatible with gauge invariance. The gauge boson, the photon, must therefore be massless.

In (12.2) we are changing the phase of the electron wavefunction locally and the gauge field, A_μ, has to be introduced in such a way as to compensate for the effect of phase differences which would otherwise be observable. The gauge field must have infinite range (that is the photon must be massless), since there is no limit to the distance over which the phases of the electron field might have to be reconciled.

Charge-changing weak interactions involve transitions of the type $u \leftrightarrow d$ and/or $e^- \leftrightarrow \nu_e$. For example, the quark-level description of β decay $(n \rightarrow pe^- \bar{\nu}_e)$ is $d \rightarrow ue^- \bar{\nu}_e$, while that of π^- decay $(\pi^- \rightarrow \mu^- \bar{\nu}_\mu)$ is $\bar{u}d \rightarrow \mu^- \bar{\nu}_\mu$. Thus for weak interactions it is natural to try to extend the U(1) local gauge symmetry of (12.2) to an SU(2) isospin-like gauge group, with invariance under local gauge transformations such as

$$\begin{pmatrix} u \\ d \end{pmatrix}_L \rightarrow e^{i\phi_\alpha(x)\tau^\alpha} \begin{pmatrix} u \\ d \end{pmatrix}_L \qquad (12.6)$$

where u and d denote the quark wavefunctions and τ^α, with $\alpha = 1, 2, 3$, are the Pauli matrices. A similar transformation exists for the lepton doublets $(\nu_e, e^-)_L$ etc. The subscript L is to remind us that only transitions between left-handed fermions (or right-handed antifermions) occur in charge-changing weak interactions. Non-Abelian gauge symmetries of this type were first considered by Yang and Mills (1954) and have the special property of leading to renormalisable field theories.

Proceeding in analogy to QED, we find invariance under (12.6) can be achieved if we introduce a covariant derivative

$$D_\mu = \partial_\mu + ig\tfrac{1}{2}\boldsymbol{\tau} \cdot \boldsymbol{W}_\mu \tag{12.7}$$

which contains three gauge fields W_μ^α with $\alpha = 1, 2, 3$. The gauge fields correspond to vector (spin-1) bosons which couple with strength g in a specified way to (left-handed) lepton and quark SU(2) doublets. The right-handed fermions are assigned to SU(2) singlets and so do not couple to W^α. Similarly the SU(3) colour theory of the strong interaction, QCD, involves colour triplets of quarks which couple to an octet of massless gluons (cf figure 3).

In table 4 we compare the structure of the gauge symmetries found in particle physics. In addition to the basic vertices shown in the table, the non-Abelian character of the QCD and weak interaction symmetries leads to a self-coupling between the gauge bosons (see, for example, the three-gluon vertices in figure 4(b)) whose structure is also determined by the gauge group.

Table 4. Gauge symmetries of particle physics.

	QED	QCD	Weak†
gauge group	$U(1)_Q$	$SU(3)_{colour}$	$SU(2)_L$
basic vertex			
vertex coupling	$-ieQ_f$	$-ig_s \dfrac{\lambda_{ij}^\alpha}{2}$ $\alpha = 1, \ldots, 8;\ i, j = 1, 2, 3$	$-ig\dfrac{\tau_{ij}^\alpha}{2}$ $\alpha = 1, 2, 3;\ i, j = 1, 2$
gauge bosons	photon A	8 gluons $G^\alpha;\ \alpha = 1, \ldots, 8$	3 weak bosons $W^\alpha;\ \alpha = 1, 2, 3$
fermions f	singlets $e^-, \ldots, Q = -1$ $u, \ldots, Q = \tfrac{2}{3}$ $d, \ldots, Q = -\tfrac{1}{3}$	triplets of colour $i, j = R, G, B$ $\begin{pmatrix} u^R \\ u^B \\ u^G \end{pmatrix}, \begin{pmatrix} d^R \\ d^B \\ d^G \end{pmatrix}, \ldots$	left-handed isospin doublets $\begin{pmatrix} \nu^e \\ e^- \end{pmatrix}_L, \ldots, \begin{pmatrix} u \\ d \end{pmatrix}_L, \ldots$ right-handed singlets $e_R^-, u_R, d_R, \ldots$

group structure of vertices:

$$-ie\langle f|Q|f\rangle A \qquad -ig_s\langle f_i|\tfrac{1}{2}\lambda^\alpha|f_j\rangle G^\alpha \qquad -ig\langle f_i|\tfrac{1}{2}\tau^\alpha|f_j\rangle W^\alpha$$

spin structure of vertices (spin-1 bosons A_μ, G_μ, W_μ; spin-$\tfrac{1}{2}$ fermions f):

vector coupling (γ^μ)　　　　　　　　　　　　　　　　　　vector–axial vector

$\langle f|\gamma^\mu|f\rangle A_\mu = (\bar{f}_R\gamma^\mu f_R + \bar{f}_L\gamma^\mu f_L)A_\mu$　　　　$\langle f|\gamma^\mu\tfrac{1}{2}(1-\gamma^5)|f\rangle W_\mu$

(also $A_\mu \to G_\mu$)　　　　　　　　　　　　　　　　　　$= \bar{f}_L\gamma^\mu f_L W_\mu$

where $f_{R,L} \equiv \tfrac{1}{2}(1 \pm \gamma^5)u_f$ are right-/left-handed Dirac spinors

† The electroweak gauge symmetry is $SU(2)_L \times U(1)_Y$, which embodies $U(1)_Q$ of QED (see the text and, in particular, figure 121).

Actually the weak interaction is not quite as simple as shown, since it is mixed with QED through the breakdown of the gauge symmetry. This mixing is sometimes referred to as the unification of weak and electromagnetic interactions, though in fact there still remain two independent coupling constants e and g. We speak of the 'standard model' of electroweak interactions. It was originally proposed by Weinberg (1967), Salam (1968) and Glashow (1961) and is a $SU(2)_L \times U(1)_Y$ gauge symmetry, and is in impressive agreement with all observed electroweak phenomena. The $U(1)_Q$ gauge symmetry of QED is embedded in this group. Y is called the 'hypercharge' in analogy to the charge Q of $U(1)_Q$. The embedding of QED is specified by the relation

$$Q = T^3 + \tfrac{1}{2}Y \tag{12.8}$$

where the third component of the isospin $T^3 = \tfrac{1}{2}\tau^3 = \pm\tfrac{1}{2}$ for the members of an $SU(2)_L$ doublet. Equation (12.8) is mathematically equivalent to the Gell-Mann–Nishijima formula for hadronic isospin and hypercharge, so we refer to Y as the 'weak hypercharge' and speak of $SU(2)_L$ as 'weak isospin'.

Local gauge invariance would require all the gauge bosons to be massless, and yet the vector bosons mediating the *short-range* weak interactions are massive. These masses are generated by breaking the gauge symmetry in a rather special way. To be precise, the Lagrangian retains the gauge symmetry, but the symmetry is 'hidden' (or, as we say, 'spontaneously broken') in the ground state of the system. It is possible to show that the theory remains renormalisable, though the proof is difficult and took several years, and we shall not go into the details here. In its simplest form the masses are generated at the expense of introducing a symmetry breaking neutral, spin-0 field called the Higgs field (for introductory accounts see, for example, Fritzsch and Minkowski 1981, Okun 1982, Aitchison and Hey 1982, Halzen and Martin 1984). The mass of the associated 'Higgs' particle, H, is not specified, but for the theory to be acceptable it is expected to lie somewhere in the range $10\,\mathrm{GeV} \lesssim m_H \lesssim 1\,\mathrm{TeV}$ (Ellis *et al* 1976a). Its coupling to fermions (and bosons) is proportional to their masses, that is to say

$$g(H f \bar{f}) \propto g \frac{m_f}{M_W}$$
$$g(H W \bar{W}) \propto g M_W \tag{12.9}$$

where g is the gauge coupling of (12.7). Since only light fermions, with $m_f \ll M_W$, are experimentally abundant, the smallness of the coupling means that the Higgs boson is rarely produced in particle interactions and so will be hard to detect.

In the standard electroweak model we denote the four gauge bosons associated with the $SU(2)_L$ and $U(1)_Y$ symmetries by $W^\alpha (\alpha = 1, 2, 3)$ and B respectively. Then the charged weak bosons are $W^\pm = (W^1 \mp i W^2)/\sqrt{2}$, whereas the physical neutral weak boson (Z^0) and the photon (A) are the orthogonal linear combinations of the neutral gauge bosons W^3 and B,

$$Z^0 = W^3 \cos\theta_W - B \sin\theta_W$$
$$A = W^3 \sin\theta_W + B \cos\theta_W \tag{12.10}$$

where θ_W is the electroweak mixing angle (or Weinberg angle). One consequence of the electroweak mixing is that the Z^0 boson, unlike $W^\pm$, acquires some coupling to right-handed fermions. Another consequence is that $M(Z^0) \neq M(W^\pm)$.

Now in the minimal (or Weinberg–Salam) model, the weak boson masses are generated by the introduction of just a single Higgs $SU(2)_L$ doublet. By requiring the photon (A) to be massless and to have coupling e to charged fermions, it follows that

$$e = g \sin \theta_W \tag{12.11}$$

where g is the $SU(2)_L$ coupling introduced in table 4. The electroweak mixing angle θ_W is determined by experiments involving Z^0 exchange, which are called weak neutral-current interactions, such as $\nu_\mu e^- \to \nu_\mu e^-$ and $\nu_\mu N \to \nu_\mu X$. These, and other, experiments give $\sin^2 \theta_W \approx 0.225$. Using this value, together with (12.11), the Weinberg–Salam model predicts

$$M(W^\pm) = \left(\frac{\sqrt{2}}{8G}\right)^{1/2} g = \left(\frac{\pi\alpha}{\sqrt{2}G \sin^2 \theta_W}\right)^{1/2} = 79 \text{ GeV}$$

$$M(Z^0) = M(W^\pm)/\cos \theta_W = 89 \text{ GeV} \tag{12.12}$$

where $G = 1.17 \times 10^{-5}$ GeV^{-2} is Fermi's weak coupling constant. Higher order electroweak radiative corrections shift these values upwards by approximately 4 GeV, and also complicate the determination of θ_W from low-energy weak interaction phenomena (Marciano and Sirlin 1981, Wheater and Llewellyn Smith 1982). In summary, the overall agreement of the predictions of the standard electroweak model with experiment (cf (12.1)) is extremely impressive.

The explicit vertex factors in the standard electroweak model are shown in figure 121. The Z coupling to a fermion is a mixture of the pure (left-handed) V–A coupling of $SU(2)_L$ and the vector coupling of QED, with

$$c_V = T^3 - 2 \sin^2 \theta_W Q$$

$$c_A = T^3 \tag{12.13}$$

where T^3 is the third component of weak isospin. The couplings to the various leptons and quarks are listed in table 5.

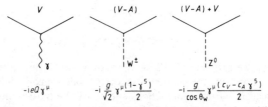

Figure 121. The gauge boson–fermion vertex factors in the standard electroweak model; $g = e/\sin \theta_W$. The vector (γ^μ) and axial-vector ($\gamma^\mu \gamma^5$) nature of the couplings is indicated. See also table 4.

Table 5. The $Z \to f\bar{f}$ couplings in the standard electroweak model (with $\sin^2 \theta_W = 0.225$).

f	T^3	Q	c_A	c_V
ν_e, ν_μ, ν_τ	$\frac{1}{2}$	0	$\frac{1}{2}$	$\frac{1}{2}$
e^-, μ^-, τ^-	$-\frac{1}{2}$	-1	$-\frac{1}{2}$	$-\frac{1}{2} + 2 \sin^2 \theta_W \approx -0.05$
u, c, t	$\frac{1}{2}$	$\frac{2}{3}$	$\frac{1}{2}$	$\frac{1}{2} - \frac{4}{3} \sin^2 \theta_W \approx 0.20$
d, s, b	$-\frac{1}{2}$	$-\frac{1}{3}$	$-\frac{1}{2}$	$-\frac{1}{2} + \frac{2}{3} \sin^2 \theta_W \approx -0.35$

A straightforward calculation gives the total width of the $W^{\pm}$ (and also, as it happens, of the Z^0) to be $\Gamma \approx 2.5 \, \text{GeV}$, increased to about $2.9 \, \text{GeV}$ by electroweak radiative corrections. The predicted branching ratios of the various W and Z decay modes are listed in table 6. From figure 121 we see that W bosons couple universally to all fermion–antifermion pairs. Thus, with 3 generations of leptons and quarks, and with 3 colours of each quark flavour, we have branching fraction $\frac{1}{12}$ for each decay mode.

Table 6. Branching ratios of the W^+ and Z^0 decay modes.

$W^+ \to e^+ \nu_e (\mu^+ \nu_\mu, \tau^+ \nu_\tau)$:	0.08	$Z \to e^+ e^- (\mu^+ \mu^-, \tau^+ \tau^-)$:	0.03
		$Z \to \bar{\nu}_e \nu_e (\bar{\nu}_\mu \nu_\mu, \bar{\nu}_\tau \nu_\tau)$:	0.06
$W^+ \to \bar{d}u (\bar{s}c, \bar{b}t)$:	0.25	$Z \to \bar{d}d (\bar{s}s, \bar{b}b)$:	0.14
		$Z \to \bar{u}u (\bar{c}c, \bar{t}t)$:	0.11

The W couplings are universal (so with 3 colours of quark the branching fractions are 3 times that into leptons). The Z branching fractions are proportional to $c_V^2 + c_A^2$ (cf table 5). No allowance has been made for the kinematic suppression of the decay modes involving the massive t quark.

We need to introduce one final complication to complete our brief survey of weak interactions. The quark states produced by weak interactions are not precisely the mass eigenstates corresponding to quarks of a definite flavour. The weak bosons couple to

$$\begin{pmatrix} u \\ d' \end{pmatrix}, \begin{pmatrix} c \\ s' \end{pmatrix}, \begin{pmatrix} t \\ b' \end{pmatrix} \tag{12.14}$$

where d', s' and b' are admixtures of the quark mass eigenstates d, s and b. So the charge-changing weak interaction $u \leftrightarrow d$ should be extended to allow for transitions of the type $u \leftrightarrow s$ and $u \leftrightarrow b$ as well. We thus have to introduce a unitary 3×3 mixing matrix, U, whose elements give the relative probability amplitudes for these transitions. It is referred to as the Kobayashi–Maskawa matrix and is specified by 3 real parameters and one phase angle (see, for example, Okun (1982) or Halzen and Martin (1984) for introductory accounts). Using an obvious notation, for six flavours of quark the *magnitudes* of the elements are experimentally found to be

$$\begin{pmatrix} U_{ud} = 0.973 & U_{us} = 0.23 & U_{ub} \approx 0 \\ U_{cd} \approx 0.23 & U_{cs} \approx 0.97 & U_{cb} \approx 0.05 \\ U_{td} \approx 0 & U_{ts} \approx 0.05 & U_{tb} \approx 1 \end{pmatrix} \tag{12.15}$$

see Kleinknecht and Renk (1983); an earlier, comprehensive, discussion is given by Chau (1983).

When only two generations were known, the quark mixing was described by a 2×2 unitary matrix containing just one independent parameter, known as the Cabibbo angle θ_C,

$$U_{ud} = U_{cs} = \cos \theta_C \approx 0.97$$
$$U_{us} = -U_{cd} = \sin \theta_C \approx 0.23. \tag{12.16}$$

We notice that the matrix (12.15) is almost diagonal, which means that there is little coupling across the generations. This implies that there should be spectacular experimental signatures from the favoured cascade decays of hadrons containing heavy quarks. For example, the semileptonic decay cascade of a heavy T hadron which contains a t quark will be characterised by the emission of multiple leptons: for instance $t \rightarrow be^+\nu$ followed by $b \rightarrow ce^-\bar{\nu}$ and then $c \rightarrow se^+\nu$.

The neutrino production of charmed particles is an instructive process to study, as it draws together several different topics that we have previously discussed. The experiment consists in observing opposite-sign dimuon events arising from either νN or $\bar{\nu}$N interactions, such as

$$\nu N \rightarrow \mu^- cX$$
$$\downarrow$$
$$\rightarrow s\mu^+\nu.$$

That is, charmed particles are detected through their semileptonic decay which contributes a muon of opposite charge to the primary muon. The dominant subprocesses are given in figure 122 and some data are shown in figure 123. For $\bar{\nu}$N production

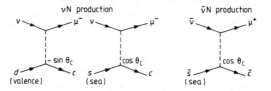

Figure 122. The production of charmed particles by neutrinos.

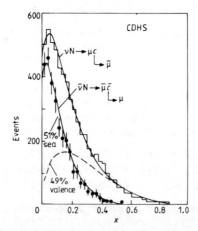

Figure 123. The x distribution of neutrino-produced opposite-sign dimuon events as observed by the CDHS collaboration at CERN.

the charmed quark can only come from a sea quark, and the $\bar{s}$ quark is favoured by the mixing element $U_{cs} \simeq \cos\theta_C$. On the other hand in νN production, transitions from both valence (with $U_{cd} \simeq -\sin\theta_C$) and sea quarks are possible, and in fact give approximately equal contributions. The Bjorken x (defined by (3.6), see also (3.10)) dependence of the data (figure 123) clearly reflects this difference, and can be subtracted to display the valence, as well as the sea, quark distributions. In summary, these data determine not only the s quark component of the nucleon, but also the U_{cd} and U_{cs} matrix elements, and the shape of the fragmentation function of the charmed quark. The latter is found to be a 'hard' distribution with $\langle z \rangle \approx 0.7$, as previously mentioned in § 11.4.

12.2   The detection of W and Z bosons at the p̄p Collider

Provided there is sufficient energy available, W and Z bosons can be produced in hadronic collisions via the Drell–Yan mechanism of figure 26. The diagram for a W^+ boson produced by p̄p collisions is shown in figure 124. However, to observe a W or

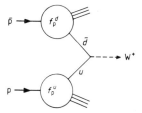

Figure 124. The Drell–Yan mechanism for W^+ production, showing only the valence quark contribution.

Z boson produced at the p̄p Collider amidst a multitude of purely hadronic events appears to be an extremely difficult task, although this was the original motivation for converting the CERN Super Proton Synchrotron into a collider (Rubbia *et al* 1976). Clearly a very distinctive signature must be sought. It is provided by isolated energetic electrons (or muons) from the two-body weak boson decays

$$\bar{p}p \to W^{\pm}X \to e^{\pm}\nu X$$
$$\bar{p}p \to Z^0X \to e^+e^-X. \tag{12.17}$$

The W boson process has the larger rate and was discovered first (UA1 collaboration 1983d, UA2 collaboration 1983b). The evidence for the $W \to e\nu$ decay is an isolated electron with a large component of momentum, p_{eT}, in the plane transverse to the beam axis. The events are also characterised by a large missing momentum, arising from the emitted neutrino which cannot be observed directly, opposite to the electron in the transverse plane: see figure 125(a). If the detector has complete solid-angle coverage this missing transverse momentum can be identified as an imbalance in the total transverse momentum. The UA1 detector at the CERN p̄p Collider has the advantage of going to within 0.2° of the beam directions, whereas the UA2 detector has coverage only to within 20° of the beams. Another distinctive feature of the two-body $W \to e\nu$ decays is the so-called 'Jacobian peak' in the electron transverse momentum distribution near $p_{eT} = \frac{1}{2}M_W$, where M_W is the mass of the W boson. The origin of this Jacobian peak is most easily seen by considering W decays at rest, figure 125(b). The transverse momentum distribution is given by

$$\frac{d\sigma}{dp_{eT}} = \frac{d\sigma}{d\cos\theta}\frac{d\cos\theta}{dp_{eT}} = \frac{d\sigma}{d\cos\theta}\left(\frac{2p_{eT}}{M_W}\right)\left(\tfrac{1}{4}M_W^2 - p_{eT}^2\right)^{-1/2} \tag{12.18}$$

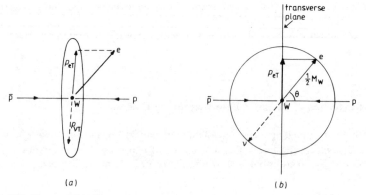

Figure 125. The two-body $W \rightarrow e\nu$ decay assuming that the W is produced on beam axis; (a) shows that the e and ν are back-to-back in the transverse plane, and (b) gives the variables if the W is at rest and the electron mass is neglected.

which follows directly from $p_{eT} = \frac{1}{2}M_W \sin \theta$. Of course the peak will be smeared by the width and transverse momentum of the W. Figure 126 shows the correlation of the transverse momenta of the electron and neutrino for 43 $W \rightarrow e\nu$ events observed with the UA1 detector. It can be seen that the events cluster about a Jacobian peak in the region $p_{eT} \simeq p_{\nu T} \simeq \frac{1}{2}M_W \simeq 40$ GeV.

A method to determine M_W, which is insensitive to the transverse motion of the W, uses a new variable, the 'transverse mass', M_T, defined by

$$M_T^2 \equiv (E_{eT} + E_{\nu T})^2 - (\boldsymbol{p}_{eT} + \boldsymbol{p}_{\nu T})^2$$
$$\simeq 2p_{eT}p_{\nu T}(1 - \cos \phi) \tag{12.19}$$

where $E_{iT}^2 \equiv p_{iT}^2 + m_i^2$, and ϕ is defined in figure 126. For $W \rightarrow e\nu$ decays it can easily be shown that M_T is essentially independent of the W momentum, satisfies $0 \leqslant M_T \leqslant M_W$, and has a distribution with a Jacobian peak near $M_T = M_W$ (Barger *et al* 1983c).

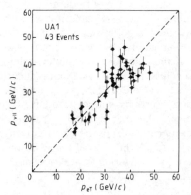
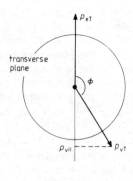

Figure 126. The correlation between the electron and neutrino transverse momenta for $W \rightarrow e\nu$ decay events (UA1 collaboration 1983e). The variables are defined in the right-hand diagram.

The observed cross section for the production of a W followed by its $e\nu$ decay, at the collider energy $\sqrt{s} = 540\,\text{GeV}$, is (UA1 collaboration 1983e)

$$\sigma(\bar{p}p \to WX \to e\nu X) = 0.53 \pm 0.08\,\text{nb} \tag{12.20}$$

which is only about 10^{-8} of the total $\bar{p}p$ cross section of roughly 70 mb! The cross section for Z production and decay into lepton pairs is about an order of magnitude smaller than that for W (see (12.25) below). The UA1 and UA2 collaborations (1983f, 1983c) detected a total of some ten Z events. The observed masses of the bosons are given in (12.1).

The W and Z production cross sections can be calculated using the Drell–Yan mechanism. The valence quark contribution to W^+ production was shown in figure 124. In analogy to (3.11), we have

$$\sigma(\bar{p}p \to W^+ X) = \tfrac{1}{3} \sum_{q,q'} \int_0^1 dx_a \int_0^1 dx_b [f_{\bar{p}}^{\bar{q}}(x_a) f_p^{q'}(x_b)$$
$$+ f_{\bar{p}}^{q'}(x_a) f_p^{\bar{q}}(x_b)] R(\bar{q}q' \to W^+) \delta(x_a x_b s - M_W^2) \tag{12.21}$$

where $\tfrac{1}{3}$ is due to colour. The subprocess cross section, $\hat{\sigma}$, is the product $R\delta(\hat{s} - M_W^2)$ in (12.21), R being the dimensionless square of the $\bar{q}q'W$ coupling. Taking the (universal) standard model W couplings of figure 121 the subprocess cross sections are all given by

$$R(\bar{q}q' \to W) = \frac{\pi g^2}{4} |U_{qq'}|^2 = \frac{\pi^2 \alpha}{\sin^2 \theta_W} |U_{qq'}|^2 \tag{12.22}$$

using (12.11). Since only left-handed quarks and right-handed antiquarks contribute, the two terms in (12.21) correspond to the production of a W^+ boson of spin projection $+1$ or -1 along the antiproton beam direction respectively, and so lead to different angular distributions of the positron emitted in the $W^+ \to e^+ \nu$ decay, see figure 127.

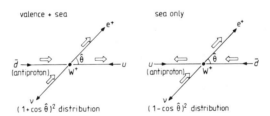

Figure 127. W^+ production and decay in the W rest frame. Other flavours of sea quark will contribute. The superimposed broad arrows represent the particle helicities (the fermion masses are neglected).

The subprocess $\bar{d}u \to W^+$ of figure 124 is dominant as it is the only one which involves the annihilation of valence quarks in both the antiproton and the proton, and it gives rise to a pronounced asymmetry in the angular distribution of the decay e^+. The positron is emitted preferentially close to the antiproton beam direction. This helicity argument is identical to the one given earlier for neutrino–quark scattering (see below equation (3.10)).

The e^+ angular distribution calculated in the $\bar{p}p$ *CM frame* using valence and sea quark densities is shown in figure 128 for different intervals of the transverse momentum of the emitted e^+. For values of p_{eT} in the Jacobian peak region, the angle $\hat{\theta}$ in the

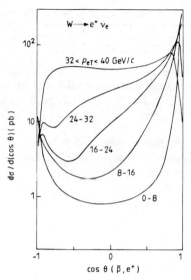

Figure 128. The e^+ angular distribution from the process $\bar{p}p \to W^+ X \to e^+ X$ for various intervals of momentum transverse to the $\bar{p}$ beam axis. The figure is from Barger *et al* (1984a).

$\bar{q}q'$ frame is, of kinematic necessity, more concentrated about $90°$ and the angular distribution simply reflects the longitudinal motion of the W. The peak in the backward e^+ direction comes entirely from sea quark annihilations.

The quark structure functions are required at x values typically in the region

$$x_a \simeq x_b \simeq \frac{M_W}{\sqrt{s}} \tag{12.23}$$

see (12.21), which is 0.15 for $\sqrt{s} = 540\,\text{GeV}$ where scaling violations of the structure functions are small. The product $f(x_a)f(x_b)$ in (12.21) is appreciable and leads to a reasonable cross section for W production, which was the primary motivation for the SPS collider (Rubbia *et al* 1976). For $\bar{p}p$ colliders at higher energies, the x values will be smaller, the sea quark contributions will be more important, the cross section will rise, and the decay asymmetry will be considerably weakened.

There are gluon corrections to (12.21). As in (4.8) the resulting leading logarithms are simply absorbed as scaling violations in $f(x, Q^2)$; but the quark densities have to be those at $Q^2 \simeq M_W^2$. The non-leading logarithms may give a multiplicative renormalisation of (12.21), known in the literature as the Drell–Yan K factor (Altarelli *et al* 1979, Parisi 1980) which has a value typically in the region of 1.5–2.

The calculation of the cross section for Z^0 production proceeds exactly as for the W. With the standard model couplings of figure 121, the subprocess cross section to be used in (12.21) is given by $\hat{\sigma} = R\delta(\hat{s} - M_Z^2)$ with

$$R(\bar{q}q \to Z^0) = \frac{\pi^2 \alpha}{\sin^2 \theta_W} \left(\frac{c_V^2 + c_A^2}{\cos^2 \theta_W} \right) \tag{12.24}$$

where $c_V^2 + c_A^2$ is equal to 0.29 and 0.37 for the $\bar{u}u \to Z^0$ and $\bar{d}d \to Z^0$ subprocesses respectively (see table 5).

Using quark densities obtained from deep inelastic scattering, evolving in Q^2 up to M_W^2 or M_Z^2, and inserting these into (12.21) together with $\hat{\sigma}$, it is found that

$$B_{e\nu}\sigma(W) = 0.3 \text{ nb}$$
$$B_{ee}\sigma(Z^0) = 0.04 \text{ nb} \tag{12.25}$$

where the branching ratios, B, are taken from table 6. The relative suppression of the predicted $Z \to e^+e^-$, as compared to the $W \to e\nu$, rate at the p̄p Collider is due to (i) the difference in branching ratios (see table 6), (ii) the fact that there are two W's and one Z, (iii) the relative couplings, that is (12.24) as compared to (12.22), and (iv) the reduction in the available phase space, and the higher x values needed, for the Z as compared to the W, due to $M_Z > M_W$. Considering that the uncertain K factor enhancement has not been included, these Drell–Yan predictions are in excellent agreement with the W and Z rates observed at the Collider (see (12.20)). Moreover, the ratio $\sigma(W)/\sigma(Z)$, which is independent of the K factor, is in excellent accord with theoretical expectations.

From a knowledge of both p_{eT} and $p_{\nu T}$ the UA1 collaboration (1983e) find the transverse momentum of the W, $p_{WT} = p_{eT} + p_{\nu T}$, has the distribution shown in figure 129(a), with an average value $\langle p_{WT} \rangle = 6.3 \text{ GeV}/c$. We comment on the theoretical implications of this in the next section.

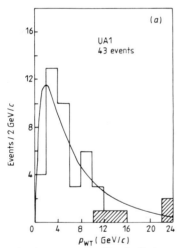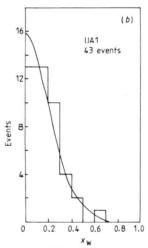

Figure 129. The transverse and longitudinal momentum distributions of W bosons produced at the CERN p̄p Collider (UA1 collaboration 1983e). The longitudinal distribution is plotted against x_W, the fractional beam energy. For the shaded events in (a) the W is accompanied by a recognisable jet. The curve in (a) is a QCD prediction (Halzen *et al* 1982a) based on (12.29).

Furthermore, if we know p_e, $p_{\nu T}$ and the W mass, then the two-body $W \to e\nu$ decay kinematics give a quadratic equation for the longitudinal component of the neutrino momentum. In the majority of the UA1 W events the two-fold ambiguity is resolved by considering energy and momentum conservation in the overall event. For the remaining events the two solutions are quite close. In this way the longitudinal momentum distribution of the W boson can be determined. The UA1 (1983e) results are displayed in figure 129(b), together with a curve showing the predictions of a Drell–Yan calculation.

Now, from a knowledge of the W momentum we can transform to the W rest frame and determine the emission angle $\hat{\theta}$ of the electron (positron) with respect to the incoming proton (antiproton). The UA1 (1983e) acceptance corrected distribution is shown in figure 130 for those events for which the charge of the decay lepton is determined. The agreement with the dominant $(1+\cos\hat{\theta})^2$ distribution (see figure 127) expected for a spin-1 W boson is impressive. Recall the analogous confirmation of the spin-1 character of the gluon discussed in § 11.2.

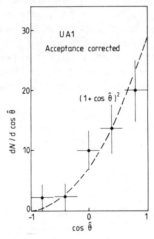

Figure 130. The angular distribution of electrons from $W \to e\nu$ decays, shown in the W rest frame (UA1 collaboration 1983e).

12.3 The W and Z as a 'collider physics workshop'

The ability to produce W and Z bosons at the $\bar{p}p$ Collider, albeit at a low rate, opens the door to the exploration of many other physics issues.

The accurate measurement of the weak boson masses and widths, together with a precise value of $\sin^2\theta_W$, can provide detailed checks of the electroweak gauge theory. Moreover since $Z \to \nu\bar{\nu}$, a measurement of the total width, Γ_Z, should reveal the total number of light neutrinos since from table 6 we see that for three families the $\nu\bar{\nu}$ modes contribute about 20% of the width.

Such precision measurements are extremely difficult for bosons produced in hadronic collisions. However there are other possibilities. For example it can be readily shown (Halzen and Mursula 1983, Hikasa 1984) that the relative rate of W and Z production is

$$\frac{\sigma(W \to e\nu)}{\sigma(Z \to ee)} \propto \frac{\Gamma_Z}{\Gamma_W}$$

with a proportionality constant which can be much more reliably calculated than can the absolute production rates.

We turn now to the interpretation of the transverse momentum distribution of the W bosons produced at the $\bar{p}p$ Collider, which we showed in figure 129(a). The events at the largest p_T values (shown shaded in the figure) have a recognisable jet recoiling against the W. These events are just what is expected from the $O(\alpha_s)$ subprocesses: $q\bar{q} \to Wg$, $gq \to Wq$ or $g\bar{q} \to W\bar{q}$. In fact gluon emission, figure 131, is the dominant subprocess. In the leading logarithm approximation the cross section is

$$\frac{1}{\sigma_0}\left(\frac{d\sigma}{d^2p_T}\right)_{O(\alpha_s)} = \frac{4\alpha_s}{3\pi^2}\frac{1}{p_T^2}\log\left(\frac{p_{T\,max}^2}{p_T^2}\right) \qquad (12.26)$$

Figure 131. W production accompanied by single gluon emission.

where $p_{T\,max}$ is the maximum allowed value of p_T, but it could equally well be replaced by any other quantity of order M_W. σ_0 is the Drell–Yan cross section for W production. Equation (12.26) is only a reasonable approximation at large p_T. For $p_T < p_{T\,max}$ the logarithm becomes large and compensates for α_s being small. We are thus led to consider multi-gluon emission diagrams, figure 132. It is found that every diagram with an additional factor of α_s is accompanied by a factor $\log^2(p_{T\,max}^2/p_T^2)$ and so a summation to all orders in α_s is required. Such a summation of the leading contributions is familiar in QED (see, for example, Lifshitz and Pitayevski 1976). The result is

$$\frac{1}{\sigma_0}\frac{d\sigma}{d^2p_T} = \frac{1}{\sigma_0}\left(\frac{d\sigma}{d^2p_T}\right)_{O(\alpha_s)}\exp\left[-\frac{2\alpha_s}{3\pi}\log^2\left(\frac{p_{T\,max}^2}{p_T^2}\right)\right] \qquad (12.27)$$

where in $O(\alpha_s^n)$ the boson is balanced by just one of the n emitted gluons, while the other $n-1$ gluons are soft.

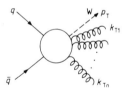

Figure 132. W production accompanied by the emission of n gluons.

The validity of (12.27) is questionable for small p_T because the exponential ('Sudakov') form factor makes the cross section $d\sigma/d^2p_T$ vanish as $p_T \to 0$. The leading logarithm approximation forces the emitted gluons to have $k_T < p_T$ even though there is plenty of phase space available for gluon emission. Consequently non-leading terms may dominate in the low p_T region and, indeed, this is what happens (Parisi and Petronzio 1979, Rakow and Webber 1981). In this region the dominant configurations have two or more gluons with larger transverse momenta which add vectorially to give the relatively small p_T of the W. These gluons are sufficiently 'hard' that the cross section can be calculated in perturbation theory.

We give a brief sketch of how this is done. As in QED, we assume the cross section for producing a W together with the emission of n gluons factorises into the product of the cross sections for emitting the gluons individually

$$\frac{1}{\sigma_0}\frac{d\sigma}{d^2 p_T} = \sum_n \frac{1}{n!} \int \frac{d^2 k_{T1}}{\pi k_{T1}^2} \cdots \int \frac{d^2 k_{Tn}}{\pi k_{Tn}^2} \left(\frac{4\alpha_s}{3\pi}\right)^n \log\left(\frac{k_{T\,max}^2}{k_{T1}^2}\right) \cdots$$

$$\times \log\left(\frac{k_{T\,max}^2}{k_{Tn}^2}\right) \delta^{(2)}(k_{T1} + \ldots + k_{Tn} + p_T). \tag{12.28}$$

By transforming to the variable conjugate to k_T (the 'impact parameter' b) the delta function can be written in a factorised form

$$\delta^{(2)}(k_{T1} + \ldots + k_{Tn} + p_T) = \frac{1}{(2\pi)^2} \int d^2 b \exp[i(k_{T1} + \ldots + k_{Tn} + p_T) \cdot b]$$

and the multigluon summation carried out. It is found that

$$\frac{1}{\sigma_0}\frac{d\sigma}{dp_T^2} = \frac{1}{4\pi} \int d^2 b \, e^{ip_T \cdot b} e^{\chi(b)} \tag{12.29}$$

where

$$\chi(b) = \int \frac{d^2 k_T}{k_T^2} \left(\frac{4\alpha_s(k_T^2)}{3\pi}\right) \log\left(\frac{k_{T\,max}^2}{k_T^2}\right) (e^{ik_T \cdot b} - 1) \tag{12.30}$$

and where the -1 takes care of the diagrams containing virtual gluons. The crucial observation is that the important region of integration in (12.30) satisfies $\Lambda^2 \ll k_T^2 \ll M_W^2$ so that the impact parameter resummation makes sense. The results of such a QCD prediction of the p_T distribution of the W are compared with the recent data in figure 129(a). Although the agreement is good it remains to see the effect of the next-to-leading logarithm contributions and of imposing longitudinal, as well as transverse, momentum conservation (Mueller 1981, Ellis *et al* 1981).

One consequence of multigluon emission is that W or Z production should be accompanied by more hadronic transverse energy (arising from the gluons) than is found in normal events. A prediction is shown in figure 133. It is clear that data on W and Z production provide a means of testing QCD techniques for calculating Drell–Yan-type processes.

Finally, we note that the W and Z bosons can be an important tool in the search for possible new particles. Examples are the top quark, t, a new heavy lepton doublet, (L, ν_L), the Higgs boson, H, or even supersymmetric partners of the existing particles. Supersymmetry postulates that fermions and bosons occur in the same multiplets, and requires the known fermions (and bosons) to have boson (and fermion) supersymmetric partners (Fayet and Ferrara 1977). If the supersymmetry were unbroken these partners would be degenerate in mass. Evidently the supersymmetry (if it exists) must therefore be broken, but it is not known by how much. A major motivation for studying supersymmetry is the search for a renormalisable field theory of the gravitational interaction (van Nieuwenhuizen 1981).

All these new particles could appear amongst the decay products of a weak boson if their mass is less than that of the boson. Possible signatures from W-initiated decays

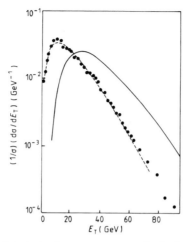

Figure 133. A prediction of the hadronic transverse energy distribution in $\bar{p}p$ events ($\sqrt{s} = 540$ GeV) in which a weak boson is produced (full curve) compared with the observed distribution of normal events (Halzen *et al* 1982b, see also Odorico 1983).

at the $\bar{p}p$ Collider are

$$W \to t\bar{b} \qquad \text{with } t \to be\nu \qquad (12.31)$$

$$W \to L\bar{\nu}_L \qquad \text{with } L \to \bar{q}Q\nu_L \qquad (12.32)$$

$$W \to WH \qquad \text{with } W \to e\nu \qquad (12.33)$$

$$W \to \tilde{W}\tilde{\gamma} \qquad \text{with } \tilde{W} \to \bar{q}Q\tilde{\gamma} \qquad (12.34)$$

where the wino $\tilde{W}$ and the photino $\tilde{\gamma}$ are the proposed supersymmetric spin-$\frac{1}{2}$ partners of the W and γ respectively.

A top quark event, (12.31), would have the distinctive signature of an isolated energetic electron recoiling against a narrow energetic $\bar{b}$ jet (Barger *et al* 1983b). The transverse momentum distribution of the jet would have a Jacobian peak near $p_T(\text{jet}) = (M_W^2 - m_t^2)/2M_W$. The event rate would be

$$\frac{\sigma(W \to t\bar{b} \to e)}{\sigma(W \to e\nu)} \sim \frac{1}{10} \times 3$$

where the 3 is for colour (see table 5) and the 0.1 is the semileptonic branching fraction for t decay. Selection criteria to eliminate background contributions will reduce the useful event rate by another factor of 3 or so. More details of this and other t production processes are given in § 13.2.

The sequential leptonic decay of a heavy lepton $W \to L\bar{\nu}_L \to e\bar{\nu}_e\nu_L\bar{\nu}_L$ has the same signature as the direct $W \to e\nu$ and the $W \to \tau\bar{\nu}_\tau \to e\bar{\nu}_e\nu_\tau\bar{\nu}_\tau$ events and so the L is hard to identify in this way. However the hadronic decay (12.32) has a promising signature of a large missing transverse momentum (due to $\bar{\nu}_L\nu_L$) balanced by two quark jets (Barger *et al* 1984a). After imposing cuts to remove background, the useful event rate is

$$\frac{\sigma(W \to L\bar{\nu}_L \to \bar{q}Q\nu_L\bar{\nu}_L)}{\sigma(W \to e\nu)} \sim \frac{1}{10}$$

for a lepton of mass 40 GeV. Supersymmetric particles can lead to similar signatures (see, for example, Ellis *et al* 1983, Chamseddine *et al* 1983, Dicus *et al* 1984).

The Higgs boson, H, will be very difficult to identify using (12.33). In this case the event rate is typically (see, for example, Finjord *et al* 1979, Keung 1982)

$$\frac{\sigma(W \to He\nu)}{\sigma(W \to e\nu)} \lesssim 10^{-3}.$$

It is clear that the existence, or otherwise, of the above particles has profound implications. It is hoped that before too long p̄p Collider data will lead either to the discovery of some of these particles or to the certainty that they do not exist, at least not with masses much below that of the W and Z bosons.

There is, of course, always the possibility that something completely unexpected may be found. An early hint that this may be so is provided by the large proportion of $Z \to l^+l^-\gamma$ events (with a hard photon) relative to $Z \to l^+l^-$ where $l = e$ or μ (Rubbia 1983, UA2 collaboration 1983c). Also various unusual events have been reported (Rubbia 1984) in which an energetic photon, or a single jet, recoils against a very large missing transverse energy, $E_T(\text{missing}) \approx 50$ GeV, which could perhaps be due to neutrinos. Moreover, other events apparently show the production of a weak boson unexpectedly accompanied by multiple energetic jets. The interpretation of all these events is, at present, obscure.

13

HEAVY QUARK PRODUCTION

In this chapter we discuss the production of particles containing heavy quarks in hadronic collisions. By heavy quark we mean charm, c, or more massive flavours of quark (b, t), which have $m_q \gg \Lambda$, the QCD mass scale, and hence are not so readily produced as states containing only light quarks. Although hadronic collisions which produce such states are comparatively rare, these events are potentially very interesting and can reveal new aspects of the way in which partons interact.

Before the advent of QCD, most models for particle production predicted an exponentially low probability, P, for the yields of particles of high mass m. For instance, in the thermodynamic statistical model (see, for example, Peterson 1979)

$$P \sim e^{-2mc^2/kT} \tag{13.1}$$

where T, the universal temperature, is equivalent to an energy of about 160 MeV. If we apply this model to particles containing (u, d), s, c and b quarks we obtain roughly

$$\pi : K : D : B \simeq 1 : 0.04 : 10^{-6} : 10^{-15}.$$

The small estimates for particles containing heavy quarks arise from the difficulty of localising enough energy for the production of a heavy particle.

This dramatic exponential suppression is absent in QCD because the coupling of quarks to gluons is independent of their flavour. The experimental observations are sketched in figure 134, and give strong qualitative support for the QCD approach. The cross section for charm production is predicted to be more than 1 mb at the $\bar{p}p$ Collider. It is however quite difficult to identify the heavy quark production events among the

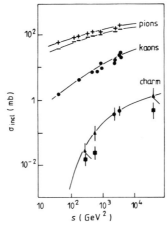

Figure 134. The total inclusive cross sections for $\pi^\pm$, K amd charm particles in pp collisions taken from Halzen (1982). The charm cross sections depend on the A^α dependence assumed in the experiments using nuclear targets ($\alpha = \frac{2}{3}$ for the triangles, $\alpha = 1$ for the squares).

much more copious light quark events. In contrast, at high energy e^+e^- colliders, although the event rate is much lower, the production of each flavour of quark depends just on e_q^2 (see (5.10)) so the background problems are much less severe. For example, at $\sqrt{s} = 30$ GeV $\sigma(e^+e^- \to q\bar{q}) \simeq 2e_q^2 \times 10^{-4}$ mb, which gives $\sigma(c\bar{c}) \simeq 10^{-4}$ mb. Of course the fragmentation of these quarks produces many more light particles, especially pions, than particles containing heavy quarks, but the proportion of heavy quark particles increases with s, as figure 134 shows, indicating that the main suppression is a threshold effect.

13.1 The hadroproduction of charm

There is a wide variety of experiments, each giving rather limited information on the hadroproduction of charmed particles through reactions of the type

$$\text{pp} \to C\bar{C}'X \qquad \pi\text{p} \to C\bar{C}'X \qquad (13.2)$$

where C (and also C′) stands for either a D meson (the lightest $c\bar{u}$ or $c\bar{d}$ bound state with $m_D = 1.86$ GeV, see table 2(a)) or a Λ_c^+ baryon (the lightest udc bound state with $m_{\Lambda_c} = 2.28$ GeV). Note that if a heavier charmed particle is produced in the collision it will decay hadronically to one of these ground states which can then be observed. As usual, X in (13.2) denotes all the other particles that may be produced. The lowest order QCD subprocesses giving rise to the reactions (13.2) are $gg \to c\bar{c}$ or $q\bar{q} \to c\bar{c}$ of figure 135. These are referred to as the 'flavour-creation' mechanism or, more simply, as 'QCD fusion'.

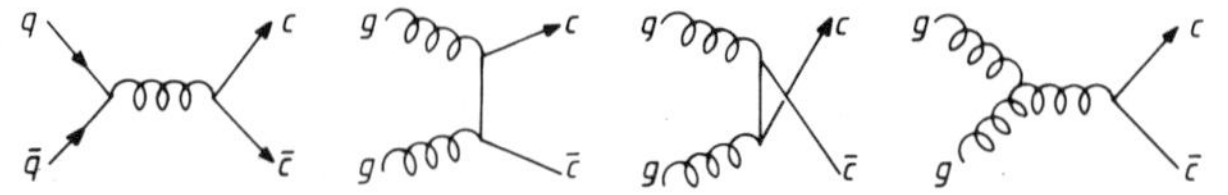

Figure 135. The lowest-order QCD flavour-creation (or fusion) subprocesses for charm hadroproduction.

The experiments may be divided into three different classes (see, for example, Muller 1983).

(i) The detection of leptons from the semileptonic decays

$$C \to \mu\nu X \qquad \text{or} \qquad C \to e\nu X. \qquad (13.3)$$

In the 'beam-dump' experiments at the CERN SPS and at FNAL, 'prompt' muons or neutrinos are detected in a small-angle forward cone after the rest of the beam, and reaction products, have been totally absorbed. This ensures that there is little contamination by non-prompt leptons from the decays of pions and kaons for example. The beam-dump experiments were among the first to detect the hadroproduction of charm and found cross sections much in excess of the QCD flavour-creation estimates based on the diagrams of figure 135. At the ISR evidence for charm production had previously been obtained by measuring the yields of large-angle decay electrons.

(ii) The observation of narrow peaks in the invariant mass of exclusive hadronic decay channels, such as

$$\Lambda_c^+ \to K^- p \pi^+.$$

ISR experiments of this type were the first to find abundant production of Λ_c^+ at large x values, shown by the squares in figure 136, again at a rate unexpected by QCD flavour-creation.

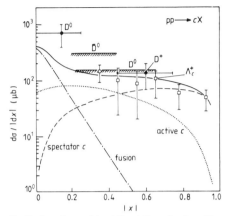

Figure 136. The $d\sigma/dx$ distribution observed for Λ_c^+ and D production. The curves are a QCD calculation (Barger *et al* 1982) based on the diagrams of figure 135 and figure 141.

(iii) The third method relies on the visual detection of the charm production and decay vertices, which are separated by a distance of order

$$l = \frac{c\tau p}{m_c} \simeq 300 \ \mu\mathrm{m} \tag{13.4}$$

for a charm particle of momentum $p = 20 \ \mathrm{GeV}/c$, with a typical lifetime $\tau = 10^{-13}$ s. Visual detectors (emulsions, high-resolution bubble chambers or streamer chambers, backed up by spectrometers to detect the secondary particles) can resolve such small distances. Although the number of events collected is small, there is essentially no background and the acceptance for $x > 0$ is almost constant. A very promising development is the active silicon microstrip target which can detect the point at which the charge multiplicity increases as a result of the decay of the charm particle downstream from the interaction point. It may be possible to design these active microvertex detectors to trigger on charm (or beauty) particle events. This would make possible high statistics experiments on heavy quark production.

A compilation of the charm production cross sections is shown in figure 137. The measurements up to $\sqrt{s} = 27 \ \mathrm{GeV}$ are obtained by fixed-target experiments at the SPS and FNAL, while the order-of-magnitude larger cross sections at $\sqrt{s} = 63 \ \mathrm{GeV}$ are measured at the ISR. The increase is interpreted as the threshold rise of charm production (see also figure 134). The dashed curve is the prediction of the QCD fusion diagrams (figure 135). There are large uncertainties in the data points. Each experiment has its own special difficulties, and usually models must be invoked to convert the measured signal into a cross section. Nevertheless there is a general consensus that QCD fusion underestimates charm production.

The dominant fusion subprocess $gg \rightarrow c\bar{c}$, for instance, gives a total charm yield

$$\sigma(\mathrm{pp} \rightarrow \mathrm{C\bar{C}X}) = \int \mathrm{d}x_a \int \mathrm{d}x_b \, f_p^g(x_a) f_p^g(x_b) \hat{\sigma}(gg \rightarrow c\bar{c}) \tag{13.5}$$

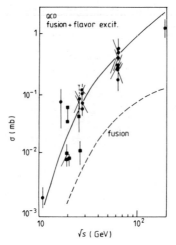

Figure 137. A compilation of data on the producton of charm in NN and πN collisions from Halzen (1982). The QCD fusion prediction could be increased in an *ad hoc* way by decreasing the mass of the charm quark (but see text). The full curve is a QCD calculation due to Odorico (1982), which is strongly dependent on a kinematic cut-off.

for $x_a x_b s > s_{th}$. The true kinematic threshold is $\sqrt{s_{th}} = 2m_D$, but $2m_c$ is more relevant if the charmed hadrons were formed in a recombination process, thereby gaining energy. Thus σ is sensitive to the choice of the quark mass m_c. It is, however, evident from a glance at figure 136 that the fusion model predicts quite the wrong x dependence, and is particularly deficient at large x. The fusion diagrams, with their $(1-x)^6$ type behaviour, only populate the 'central' (small-x) region. Some other production mechanism must be present at large x. This suggests a 'diffractive' or leading particle effect associated with the forward-góing partons as in the examples shown in figure 138. If we ignore the quark mass effects, then the x dependence as $x \to 1$ may be obtained using the counting rules of chapter 6. The predictions from (6.8) are shown on the diagrams.

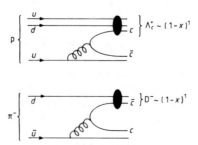

Figure 138. The counting rule predictions, (6.8) for the x behaviour of $p \to \Lambda_c^+$ and $\pi^- \to D^-$ (or D^0) forward production. From the first diagram we also predict that $p \to \bar{D}$ has a 'softer' $(1-x)^3$ behaviour. Forward production of $\pi^- \to D^+$, $\bar{D}^0$ and $p \to D$ are much more suppressed as $x \to 1$.

Figure 139, for example, shows results for D meson charm production from 360 GeV/c π^-p interactions using LEBC, the high-resolution hydrogen bubble chamber. Although the statistics are low (18 D decays), these data are particularly clean. We see that the x distribution for D$^-$ and D^0 production extends up to high x, whereas the D$^+$ and $\bar{D}^0$ mesons are all produced centrally. The difference strongly suggests a leading particle effect. A valence d quark of the π^- can recombine with a pair-created $\bar{c}$ quark to give a D$^-$ meson (see figure 138). Similarly a D^0 can be formed by $\bar{u}c$ recombination, whereas neither a D$^+$ nor a $\bar{D}^0$ can be formed in this way. In fact the data show evidence for two components of (D$^-$, D^0) production, a 'central' component with $(1-x)^6$ behaviour and a 'diffractive' component with the $(1-x)$ behaviour.

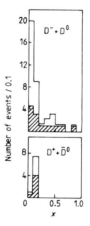

Figure 139. The acceptance corrected and uncorrected (shaded) distributions of D mesons detected in π^-p interactions using LEBC (Aguilar-Benitez *et al* 1983).

There have been several attempts to calculate large-x charm production. Here, the underlying production mechanism is much less certain than the QCD 'flavour-creation' description of central charm production. Two different mechanisms have been proposed: QCD 'flavour-excitation' of figure 140(a), and diffractive production as in figure 140(b). The possible flavour-excitation subprocesses, which were first

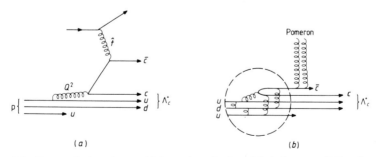

Figure 140. Two mechanisms for p → Λ_c^+ forward production. (a) A flavour-excitation diagram. (b) Diffractive production from intrinsic charm.

considered by Combridge (1979), are shown in figure 141. Combridge argues that
QCD fusion and flavour-creation are two distinct limits of the same diagram, and so
should not amount to double counting, but it is obviously difficult to be sure about this.

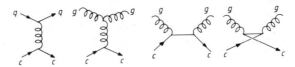

Figure 141. The QCD flavour-excitation diagrams for charm production.

Both of the large-x production mechanisms involve interactions with a charm quark
component of the hadron as in figure 140. However it is important to distinguish the
origin of the charm component in the two pictures.

In figure 140(b) the charm component exists over a time scale independent of
any probe momentum and is associated with the bound state dynamics of the proton
(as recorded by the extra gluons in the diagram). This is called the 'intrinsic' charm
component of the proton (Brodsky *et al* 1981). It is a non-perturbative effect and so
it is far from obvious how to estimate the magnitude of this component from QCD,
though it is expected that the $x \to 1$ behaviour of the intrinsic charm distribution may
be predicted by the techniques of chapter 6. In this approach we must determine or
specify the probability that the proton is found in a *uudcc̄* configuration. This probability is usually taken to be about 1% to reproduce the magnitude of the observed cross
section. We also require the $f_p^c(x)$ distribution. It must be reasonably 'hard' so that
after diffractive scattering (shown by Pomeron exchange in figure 140(b)) the c quark
can recombine with *ud* valence quarks to form a forward going Λ_c^+. This is quite
different from the $f_p^c(x)$ distribution measured in deep inelastic scattering at large Q^2
which is concentrated at small x, although the QCD evolution (sketched in figure 34)
has a tendency to shift the distribution to lower x as Q^2 increases.

We turn now to the flavour-excitation mechanism of figures 141 and 140(a). Here
the charm quark is created on a short time scale in association with a large momentum
Q^2, and so the $f_p^c(x)$ distribution may be derived from QCD bremsstrahlung and $c\bar{c}$
pair production. It satisfies the standard QCD evolution. Nevertheless this mechanism,
too, is ambiguous. For example, the contribution of diagram 140(a) contains

$$\hat{\sigma}(gc \to gc) = \int_{\hat{t}_{min}} d\hat{t} \frac{1}{\hat{t}} \cdots \tag{13.6}$$

where we have exhibited a divergence due to the gluon exchange pole. However, the
gluon plays the same role as the virtual photon in deep inelastic scattering (figure 25),
and so $-\hat{t}$ is analogous to the Q^2 specifying the QCD evolution of the structure function.
For $\hat{t} \sim 0$ no charm is present in the proton and so the divergence is absent. Only
when $\hat{t} \simeq -m_c^2$ have enough $c\bar{c}$ pairs evolved for charm production to take place. The
dominant region is $\hat{t} \sim \hat{t}_{min}$ and so we can simply evolve up to $Q_0^2 = -\hat{t}_{min} \simeq m_c^2$. The
result of the calculation is very sensitive to the choice of Q_0^2. Increasing Q_0^2 increases
$f_p^c(x)$, but dramatically reduces $\hat{\sigma}$. The most complete treatment is due to Odorico
(1982), who allows for the phase space constraints and the quark mass, m_c, in the QCD
evolution. He uses the observed transverse momentum distribution for charm production to constrain Q_0^2, since this distribution can be predicted by the QCD evolution
procedure. He finds a satisfactory description can be obtained with the choice $Q_0^2 = \frac{1}{4}m_c^2$.

His result using QCD flavour-excitation, as well as fusion, is shown by the curve on figure 137.

The flavour-excitation mechanism 'impersonates' the diffractive production of charm. For example a leading Λ_c^+ is formed by the recombination of the spectator c quark of figure 140(a) with the valence ud pair. Alternatively the $\bar{c}$ quark can recombine with a u or d valence quark to form a leading $\bar{D}$ meson, which will, on average, be produced at lower x than Λ_c as it contains only a single valence quark. A sample calculation along these lines is shown in figure 136, although it is based on a 'harder' $f_p^c(x)$ distribution than would be obtained by evolution.

In summary, forward charm production is not well understood. Flavour-excitation diagrams are perhaps able to account for the data, but are strongly dependent on the kinematic cut-off in (13.6). With increasing quark mass, flavour-excitation decreases much more rapidly than the m_Q^{-2} behaviour expected of truly diffractive processes. This has important implications for b and t quark production, a subject to which we now turn.

13.2 Heavy flavour production at the p̄p Collider

We have already discussed several mechanisms which could lead to top and bottom quark production in high-energy hadronic collisions. The various subprocesses are
(i) production via a weak boson (§ 12.3)

$$W \to t\bar{b}, \bar{t}b$$
$$Z \to b\bar{b}, t\bar{t} \tag{13.7}$$

    provided $m_t < M_W$ and $m_t < \frac{1}{2}M_Z$ respectively,
(ii) QCD flavour-creation or fusion (§ 13.1)

$$gg, q\bar{q} \to b\bar{b}, t\bar{t} \tag{13.8}$$

(iii) diffraction and QCD flavour-excitation (§ 13.1)

$$gb \to gb \quad \text{(with a spectator } \bar{b})$$
$$gt \to gt \quad \text{(with a spectator } \bar{t}) \tag{13.9}$$

and the related processes with g replaced by q, and with $t \leftrightarrow \bar{t}$ etc.

The production cross sections obtained from these subprocesses depend on the quark mass m_Q. The behaviour is shown in figure 142 for p̄p collisions at $\sqrt{s} = 540$ GeV. The prediction for weak production ($\bar{p}p \to W \to t\bar{b}, \bar{t}b$) should be reliable as it is based on the observed $W \to e\nu$ cross section, (12.20). The decrease, with increasing $m_Q = m_t$, just represents the kinematic suppression as m_t approaches M_W. There may be a little more uncertainty in the QCD fusion prediction depending on the choice of structure functions etc, but it would be surprising if it was very far wrong. However the curves for the mechanisms of (iii) are little more than physically motivated guesses. Diffractive processes are usually taken to have a m_Q^{-2} dependence, to characterise the cross section presented by a heavy quark to the incoming parton, and a threshold dependence (arbitrarily taken to be of the form $\log \sqrt{s}/m_Q$ to give an upper limit). The curve is normalised to a diffractive charm production cross section of 0.5 mb. From figure 142 we conclude that for a top quark of mass $m_t = 35$ GeV the production cross section at the collider is in the range $2 \leqslant \sigma \leqslant 100$ nb.

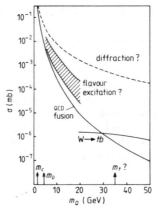

Figure 142. The cross sections for heavy quark production in $\bar{p}p$ collisions at $\sqrt{s} = 540\,\text{GeV}$, via the mechanisms of (13.7)–(13.9), as a function of the quark mass m_Q. $W \to tb$ includes both $t\bar{b}$ and $\bar{t}b$ production. The predictions for diffraction and flavour excitation are not much more than inspired guesses. The broken curve should be regarded as an upper limit for diffraction.

Mechanisms (i) and (ii) will give high-p_T t and b quark jets. However, it will be essentially impossible to identify these directly from the much, much more numerous light-quark and gluon jets. Due to its mass ($m_t > 20\,\text{GeV}$) the t quark will give a broad jet which may sometimes appear as three sub-jets from the $t \to bu\bar{d}$ decay.

The best possibility for distinguishing processes (i)–(iii) is through the semileptonic decays of the produced T and B hadrons, which at the quark level are given by

$$t \to b\mu\nu, \qquad b \to c\mu\nu, \qquad\qquad (13.10)$$

or with μ replaced by e (see Pakvasa *et al* 1979). The branching fractions of these modes are expected to be about 10% (since eν, $\mu\nu$, $\tau\nu$, $u\bar{d}$, $c\bar{s}$ occur equally and there are three colours of quark). The two heavy-quark jets arising from W decay, or from QCD fusion, will be approximately back-to-back in the plane transverse to the beam axis, since the annihilating quark and antiquark (or gg) are incident essentially along the beam axes. The different event topologies are sketched in figure 143.

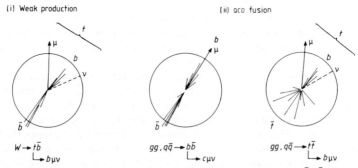

Figure 143. Idealised event topologies, in the plane transverse to the beam axis, for $t\bar{b}$, $b\bar{b}$ and $t\bar{t}$ production and subsequent semileptonic decay. Isolated energetic muons arise from t decay, whereas the decay muons from $b\bar{b}$ (and $c\bar{c}$) events will be within the accompanying hadron decay jet. The figure applies equally well to the eν decay modes. We emphasise muons since, with present detectors, they can be identified down to lower transverse momenta than electrons, allowing higher event rates.

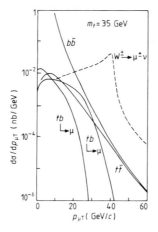

Figure 144. The predicted muon p_T distribution from heavy quark production and decay in p̄p collisions at $\sqrt{s} = 540$ GeV, with $m_t = 35$ GeV. The $t\bar{t}$ and $b\bar{b}$ curves include both QCD fusion and Z sources; $c\bar{c}$ results are similar to $b\bar{b}$ for $p_T \gtrsim 10$ GeV/c. The $W^+ \to t\bar{b}$, $b\bar{t}$ contributions (denoted tb) are separated into two parts, according to whether the muon comes from t or b decay. All rates are summed over μ^+ and μ^-. For comparison the $\mu^\pm$ spectrum from the direct $W^\pm \to \mu^\pm \nu$ decay is also shown. The figure is taken from Barger *et al* (1984b).

Figure 144 shows the transverse momentum distribution of the muons arising from $b\bar{b}$, $t\bar{t}$, $t\bar{b}$ and $\bar{t}b$ production and their subsequent semileptonic decay, (13.10). Such predictions involve the fragmentation of the heavy quark, Q, into a heavy hadron, H, which subsequently undergoes a semileptonic decay. The steps are shown in figure 145, using t decay as an example. The fragmentation functions $D_Q^H(z)$, where z is the fraction of quark momentum carried by the hadron, are expected to have increasingly 'hard' distributions as the mass of the quark increases (Suzuki 1977, Scott 1978, Bjorken 1978, Peterson *et al* 1983). Typically $\langle z \rangle \approx 0.7, 0.85$, and just below 1.0 for c, b and t fragmentation respectively, which is consistent with e^+e^- data on c and b production and decay, as we discussed in § 11.4.

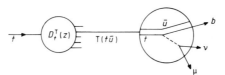

Figure 145. The fragmentation and semileptonic decay of a t quark jet. This is the first stage of the $t \to b \to c \to s$ cascade; the next stage is the fragmentation of the b jet.

For comparison, we show in figure 144 the muon distribution from $W \to \mu\nu$ events with its Jacobian peak at $p_{\mu T} \approx \frac{1}{2}M_W$, see (12.18). As anticipated by figure 142, $b\bar{b}$ (and also $c\bar{c}$) initiated events dominate the spectrum, particularly at small $p_{\mu T}$. Fortunately t quark decays have a rather distinctive property. The momentum components of the decay products perpendicular to the axis of a heavy quark jet have the kinematic limit $p_\perp < \frac{1}{2}M_Q$. Thus the muon from $b \to c\mu\nu$ decay will be contained within the hadronic debris of the accompanying c jet (see figure 143). On the other hand because the t mass is so much greater, isolated muons occur in $t \to b\mu\nu$ decays. Hence it is possible to devise selection criteria to isolate $t\bar{t}$ QCD fusion and $W \to t\bar{b}$ events, which should occur at a rate of about 10% of the $W \to \mu\nu$ signal.

The rate of diffractive and flavour-excitation production of top quarks is not known, but the cross section may conceivably be as high as 100 nb or more, see figure 142. For these events we expect an active t and a spectator $\bar{t}$ quark (or vice versa) to be produced with small p_T (figure 140). Recombination with valence quarks of the incident p (or $\bar{p}$) should then give a leading Λ_t or $\bar{T}$ ($\bar{\Lambda}_t$ or T) baryon at high x and low p_T, together with the associated production of a $\bar{T}$ or T meson at lower x. The interesting kinematic region populated by decay leptons from these events is $p_{\mu T} \approx m_t/4$ and $\theta \approx 20°$ to the beam direction (Horgan and Jacob 1981).

We have mentioned that the UA1 detector at CERN has the ability to measure the missing transverse momentum in an event, which is attributed to the emitted neutrino (or neutrinos) and denoted $\boldsymbol{p}_{\nu T}$. With such a detector there is a way to identify t quark events, from their semileptonic decay, *independent* of the production mechanism (Barger *et al* 1983a). This is to form the 'transverse mass' of the emitted muon and neutrino(s), defined by

$$M_T^2(\mu\nu) = (E_{\mu T} + E_{\nu T})^2 - (\boldsymbol{p}_{\mu T} + \boldsymbol{p}_{\nu T})^2$$

where $E_{iT}^2 \equiv m_{iT}^2 + p_{iT}^2$. The expected transverse mass distributions for the various processes are listed in table 7 and are sketched in figure 146. Thus a semileptonic B meson decay to a charmed D meson will have $M_T(\mu\nu) \leq M_B - M_D \approx 3$ GeV. So, once

Table 7. Transverse mass, $M_T(\mu\nu)$, distributions.

Decay	Range of M_T	Character of distribution
$W \to \mu\nu$	$0 < M_T < M_W$	two-body decay: distribution with a Jacobian peak at $M_T \approx M_W$
$t \to b\mu\nu$	$0 < M_T < m_t - m_b$	three-body decays: broad distribution
$b \to c\mu\nu$	$0 < M_T < m_b - m_c$	in M_T between the kinematic limits

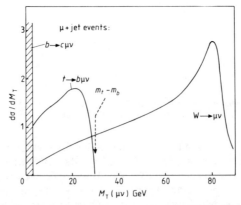

Figure 146. An idealised sketch of the expected $\mu\nu$ transverse mass distribution resulting from (i) $W \to \mu\nu$, (ii) $t \to b\mu\nu$ with $m_t = 35$ GeV, and (iii) $b \to c\mu\nu$, decays. In practice, experimental uncertainties will cause the b decay events to 'spill over' to higher M_T values; and multineutrinos from t cascade decays will give the M_T distribution a tail beyond $m_t - m_b$.

the $W \to \mu\nu$ events have been identified, the remaining large-M_T events can all be attributed to t quarks and will have an M_T distribution with a kinematic cut-off at $m_t - m_b$. Besides the large M_T, the t quark events will also be characterised by the presence of jet activity (see figure 143). Due to the $t \to b \to c \to s$ decay cascade, there will be a certain fraction of multilepton events and so $p_{\nu T}$ may be the vector sum of the transverse momenta of all the emitted neutrinos. However $p_{\nu T}$, and hence the M_T distribution, are dominated by the primary semileptonic decay.

If the b jet in the $t \to b\mu\nu$ decay is also identified and its momentum, p_b, measured, then an essentially two-body transverse mass distribution can be formed by treating the $b\mu$ system as a single cluster. This has the advantage that it has a sharp Jacobian peak at m_t (see Barger et al 1983a).

Before concluding this chapter, it is worth mentioning that 'hidden' flavour production (e.g. $pp \to \psi X$) occurs at a much lower rate than that for 'open' flavour production, such as Λ_c and D. A comparison of QCD predictions for ψ hadroproduction with data at FNAL and ISR energies has been made by Baier and Rückl (1983), based on the subprocesses $gg \to \chi_{0,2}$, $gg \to \psi g$, $gg \to \chi_J g, \ldots$, with $\chi_J \to \psi \gamma$, where χ_J are the ^3P_J ($J = 0, 1, 2$) charmonium states. At Collider energies another important mechanism is $gg \to b\bar{b} \to \psi X$, arising from the $B \to \psi X$ decay mode, so it may be possible to use the $\psi \to \mu^+\mu^-$ signal, for example, as a 'tag' for b quark production (Glover et al 1984).

In summary, it is clear that the mechanisms of heavy flavour (c, b, t quark) production are not well understood. However, rapid progress in resolving many of the outstanding problems should come from the experiments at the hadron colliders. The event rates are good, and the main problem is to isolate the heavy quark events from the much larger rates for the background processes. At present the detection of high-p_T muons (or electrons) from the semileptonic decays offers the best signature of heavy quark events. There is a promising experimental proposal which takes advantage of the relatively long lifetimes of c and b particle decays, $\tau \sim 10^{-13}$ and 10^{-12} s. By using a detector, such as a high pressure drift chamber, mounted on, or close up to, the collider beam pipe, it may be possible to identify heavy quark events by resolving the decay vertices of the decay cascade. Indeed one can foresee that heavy quarks (as well as W decay) may soon be used as a tool to study other important issues, such as the existence of the Higgs particle.

14

THE STRUCTURE OF THE HADRONS

It will be evident from the preceding chapters that our understanding of the structure of hadrons, and the mechanisms of hadron interactions, is tantalisingly incomplete. Very great progress has been made in the last twenty years or so, and we now have seemingly incontrovertible evidence that hadrons are made of more fundamental, point-like, constituents, i.e. the quarks and gluons, and that these partons are confined within hadrons. Not only does this hypothesis enable us to understand most features of the spectrum of hadrons (at least qualitatively) but the constituents are actually 'seen', almost directly, in deep inelastic scattering experiments, and the outcome of many hard-scattering experiments is the production of jets of hadrons which appear to reflect the motions of the underlying partons.

Most particle physicists now firmly believe that the SU(3) symmetric colour gauge theory, quantum chromodynamics (QCD), correctly describes the interactions of these partons. The most direct test of this hypothesis is that some features of the scattering data, such as the scaling violations at large momentum transfers, where the running coupling constant has become small, appear to be satisfactorily, if not uniquely, explained by QCD perturbation theory (Reya 1981).

Also QCD motivated potential models have been used to calculate the spectrum of heavy-quark states such as the charmonium $\psi(c\bar{c})$ states and the $\Upsilon(b\bar{b})$ states in impressive detail (Quigg and Rosner 1979, Berkelman 1983). Because these quarks are heavy, and there are a number of states below the threshold for open charm and beauty production, essentially nonrelativistic approaches based on solutions of the Schrödinger equation are not inappropriate. Indeed even for particles made entirely of the lighter u, d and s quarks, potential models are quite successful provided the quarks are given effective 'constituent' masses about $0.3\,\text{GeV}/c^2$ higher than the 'current' masses quoted in table 1 (Isgur 1978, Isgur and Karl 1978, Ono and Schoberl 1982). This extra mass presumably represents, however approximately, the colour sea of quark–antiquark pairs and gluons which each valence quark must carry round with it.

But, much more important, QCD provides a self-consistent, renormalisable, field theory which seemingly leads both to asymptotic freedom, and hence the approximate scaling of hard parton interactions, and to infrared slavery, that is the confinement of the partons within hadrons, though it must be stressed that this has yet to be proved formally (Politzer 1974).

Unfortunately, however, being able to write down the QCD Lagrangian with the interactions of table 4 does not enable one to deduce very much about the properties and interactions of hadrons from first principles. This is because the structure of a hadron depends on the interactions of the partons at low momenta (large distances) where the running coupling, α_s, is of order unity and perturbation methods are inappropriate. One important calculational technique for trying to overcome this difficulty is to curtail the infinite number of degrees of freedom enjoyed by field theories by replacing the space–time continuum with a finite lattice of space–time points (Creutz 1984). This automatically regularises the ultraviolet (short-distance) divergences, which require renormalisation if treated in perturbation theory. Clearly to have any hope of representing a hadron satisfactorily one needs a lattice spacing, a, which is very small compared to the radius of a typical hadron, R, and a lattice whose total extent, Na (N being the number of lattice sites in a given dimension) is much larger than R; that is to say we need $N \gg R/a$. The space–time lattice has N^4 lattice sites, and each site must be given the degrees of freedom of the colour field, and ideally the possibility of $q\bar{q}$ pair creation too. Hence even for the quite small lattices which have been employed to date, and with the neglect of fermion loops, a formidable amount of computer time is needed to predict the hadron spectrum and the other static properties (such as the magnetic moments, etc) of the hadrons. The results obtained so far for hadron masses seem quite promising given the comparative crudity of the approximations employed, but much work is still needed before one can be confident that really satisfactory solutions are emerging (Halliday 1983, 1984).

To try to simulate the interactions of hadrons using this sort of approach, including the creation of particles, is clearly an even more challenging problem. Some pessimists may wonder if we shall ever reach that point. After all, even though we now possess a good basic understanding of atomic theory it is still very hard to achieve accurate calculations of atomic structures; and to predict atom–atom scattering cross sections, with their complicated exchange mechanisms, is out of the question for all but the very simplest atoms.

Fortunately it is also true that a rather good understanding of soft hadron interactions has already been achieved through the exchange of particles which lie on Regge trajectories. Regge theory has successfully accounted for the bulk of high-energy hadron scattering data at small momentum transfers, for both exclusive and inclusive processes. In particular, simple Regge-pole exchange models, with exchange-degenerate, linear trajectories, and couplings which obey an SU(3)-flavour symmetry, give a very good overall description of these data with very few parameters. Of course, exchange degeneracy and SU(3)-flavour symmetry are not exact, we are not quite sure what the Pomeron trajectory really represents, and at larger momentum transfers Regge cuts become important. So to obtain a really detailed quantitative account of the data requires rather more parameters, but still remarkably few given the vast quantity of data to be explained (Kane and Seidl 1976, Irving and Worden 1977, Ganguli and Roy 1980). It thus seems clear that the strong interaction colour force between hadrons manifests itself principally through the exchange of colourless hadrons which lie on Regge trajectories. Regge theory had therefore already provided us, even before we were aware of it, with a phenomenological description of the results of the many underlying QCD processes.

Work is in progress to try to calculate large-angle hadron scattering, where perturbation theory may be satisfactory, from sums of QCD Feynman diagrams (Farrar and Neri 1983, Kanwal 1983), and to relate these results to Regge exchange at large angles (Collins and Kearney 1983). But clearly much remains to be understood. We urgently require to prove that QCD does indeed confine the partons (at least partially), and to get some idea of what these confined systems are like, and how they interact with each other.

We have also found that the standard $SU(2)_L \times U(1)_Y$ electroweak model is not only able to explain the electromagnetic- and weak-interaction properties of the hadrons through the interactions of their constituent quarks, but it has correctly predicted the masses and the production mechanisms of the newly discovered vector bosons, $W^\pm$, Z^0, which mediate the weak interaction. The quark model accounts for the many types of hadron, with flavour properties such as isospin, strangeness, charm, beauty, etc, and their various weak decays, and we are also beginning to understand the mechanisms which produce particles containing the heavier types of quarks. So, apart from the non-trivial difficulties arising from the lack of detailed calculations of the parton structure of hadrons, we now seem to have a fairly complete picture of their electroweak properties.

Despite all these successes there are many problems which remain to be solved. First there is the unexplained spectrum of quarks and leptons (table 1) whose masses exhibit no discernible pattern, and whose total number is unpredicted by current theory. Going beyond this, the fact that the QCD $SU(3)_{colour}$ and the electroweak $SU(2)_L \times U(1)_Y$ are both non-Abelian gauge theories, and the fact that quarks and leptons have such similar electromagnetic and weak interactions, strongly suggest that some more fundamental unification may be possible (Ross 1981, Langacker 1981). This could perhaps be achieved through a so-called 'grand unified gauge theory' in which $SU(3)_c \times SU(2)_L \times U(1)_Y$ is embedded. The simplest possible unifying group is $SU(5)$ whose $5^2 - 1 = 24$ gauge bosons would include not only the 8 colour gluons, the weak bosons $(W^{\pm}, Z)$ and the photon, but also 12 new bosons (X's and Y's) which have both colour and electroweak properties, and interconnect quarks and leptons through processes such as

$$qq \to X \to \bar{q}\bar{l} \tag{14.1}$$

(where q is a quark and l a lepton). Although at present energies the strengths of the colour and electroweak interactions are quite different, we know that the coupling constants, α_s and α, and the mixing parameter $\sin^2 \theta_W$, change logarithmically with energy, as in (1.6) and (1.11), and extrapolation of the measured values suggests that these couplings may become equal at $\sqrt{s} \approx 10^{15}$ GeV. It is only at this very great energy, which would correspond to the masses of the X and Y bosons, that the unification would become apparent.

Clearly such very high energies are completely unattainable by any conceivable accelerator, and it may be wondered whether there is any way in which this sort of speculation can be tested. One possibility is to note that such theories generally permit protons to decay through the agency of the heavy gauge bosons, so that for example (14.1) allows $p \to e^+ \pi^0$ through

$$p = (uud) \to Xd \to e^+(\bar{d}d) = e^+ \pi^0. \tag{14.2}$$

Of course the X, with a mass of the order of 10^{15} GeV, is highly virtual in this process, and hence the lifetime of the proton is very long. In order of magnitude it is given by

$$\tau_p = \frac{1}{m_p} \left(\frac{m_X}{m_p}\right)^4 \frac{\hbar}{c^2} \approx 10^{30} \text{ y}. \tag{14.3}$$

This is much longer than the age of the Universe ($\approx 10^{10}$ y), so very few protons will have decayed to date. But since a ton of material contains about 10^{30} nucleons, a detector containing a thousand tons might observe about one decay per day. Unfortunately such estimates are uncertain by several orders of magnitude and the experiments are very difficult. So far pioneering experiments have failed to find clearly identifiable proton decay events, suggesting that $\tau_p > 10^{31}$ y (Bionta *et al* 1983). Other forms of unification, such as those based on supersymmetry (van Nieuwenhuizen 1981) predict longer lifetimes (Ellis *et al* 1982). Regrettably the rate of background events caused by cosmic rays will make it difficult to detect proton decays if their lifetime is longer than about 10^{33} years.

Alternatively the quarks and leptons may themselves be composites of much smaller entities, which usually go under the generic name 'preons' (Lyons 1983). A simple example of a preon model is the *rishon model* (Harari and Seiberg 1981, 1982) (rishon $\equiv$ 'primary' in Hebrew), in which there are just two types of rishon: $T(3, \frac{1}{3})$ and $V(\bar{3}, 0)$, the figures in brackets being the colour representation and charge respectively, together with their antiparticles. Then a generation of fermions can be made up as follows:

$$e^+ = TTT \qquad e^- = \bar{T}\bar{T}\bar{T}$$

$$u = TTV \qquad \bar{u} = \bar{T}\bar{T}\bar{V}$$

$$\bar{d} = TVV \qquad d = \bar{T}\bar{V}\bar{V}$$

$$\bar{\nu} = VVV \qquad \nu = \bar{V}\bar{V}\bar{V}.$$

This automatically gives them the correct charge and colour, and it accounts for the equality of the magnitude of charges of the electron and proton ($= uud$). The other generations could perhaps be excited states of these configurations. Since we know that quarks and leptons are point-like down to very small distances ($<10^{-19}$ m) we need a new force, called 'hypercolour', to confine these rishons within the quarks and leptons, similar to the way in which the quarks are confined within hadrons by the colour force. Hadrons have a size

$$R_c \approx \frac{\hbar c}{\Lambda_c} \tag{14.4}$$

where Λ_c is the energy scale of colour coupling in (1.11), and we would need for the hypercolour force an energy scale $\Lambda_H > 1$ TeV $\gg \Lambda_c$ so that the rishons can be confined within a much smaller radius. Unless Λ_H is fairly close to this lower bound such theories will be very hard to test, but since a rearrangement of the rishons should permit $uu \to e^+\bar{d}$, as in (14.2), we may perhaps expect that the proton will decay with a lifetime of, in order of magnitude,

$$\tau_p \approx \frac{1}{m_p}\left(\frac{\Lambda_H}{m_p c^2}\right)^{4\,\text{or}\,8} \frac{\hbar}{c^2} \tag{14.5}$$

where the power depends on the details of the theory (Lyons 1983). The current experimental limit on the proton lifetime thus requires $\Lambda_H > 10^7$ or 10^{15} GeV depending on which power is appropriate.

If proton decay is not observed it may be that the only practical way of finding out about what happens at the very high energies invoked by the unification and composite-ness ideas will be to understand in greater detail the very early stages of the evolution of the Universe (Weinberg 1977, Steigman 1979, Dolgov and Zeldovich 1981, Ellis 1982). The success of 'hot big bang' cosmology in explaining the primordial synthesis of light elements such as deuterium, tritium, ^{3}He and ^{4}He during the first few seconds, when the energy per particle was about an MeV, has encouraged speculation about even earlier times, say 10^{-36} s, when the energy may have been as high as 10^{15} GeV per particle, and processes like that in (14.2) may have been responsible for the creation of a net positive baryon number in the Universe (Kolb and Turner 1983). Already these ideas have borne fruit in the so-called 'inflationary universe' scenario (Guth

1981, Linde 1982, Albrecht and Steinhardt 1982) which invokes the spontaneous symmetry breakdown of a grand unified theory (such as SU(5) mentioned above) in an essential way to explain many of the difficulties of the early stages of the 'big bang'.

Such speculations generally presume, however, that there is little or nothing of importance to be discovered between the energy regime explored so far (i.e. up to the W, Z masses at just under 100 GeV) and the very large energy scales of grand unification ($\simeq 10^{15}$ GeV) or compositeness ($\Lambda_H \gtrsim 10^7$ GeV). But the fact that in the past each new generation of accelerators has produced quite unexpected phenomena, should remind us of the incompleteness of our understanding of the basic structure of the Universe. There is every reason to expect that the higher energy accelerators which will be built in the near future will have similar surprises in store for us.

Hadron colliders (pp and p̄p) are planned to achieve several TeV and perhaps eventually tens of TeV, which may allow the production of Higgs bosons and even heavier quarks and leptons, if they exist. They may also enable us to discover whether quarks have an internal structure, and to check the slow (logarithmic) evolution of the basic hadron-scattering parameters, such as the total cross section and the average multiplicity, and the evolution of jets. It must be remembered, however, that as far as parton collisions are concerned the effective energy of a hadron collider is only $\bar{s} = x_a x_b s$ (see (3.3)) and that the average x is $\lesssim 0.2$ depending on the species of parton. Also many of the cross sections we are interested in will be even smaller than those for W, Z and t-quark production, down at the picobarn ($\equiv 10^{-12}$ barns $= 10^{-40}$ m^2) level, so the event rates will necessarily be very low compared to the total background of hadronic events (total cross section $\simeq 10^{-1}$ barns $= 10^{-29}$ m^2). It is therefore important that there should also be complementary e$^+$e$^-$ colliders such as the planned LEP at CERN and the Stanford Linear Collider (SLC) which will achieve energies of 100 GeV, sufficient to produce the Z boson copiously without the very large backgrounds of a hadron collider. They should help us to discover, *inter alia*, how many light neutrino species there are and whether there are Higgs bosons (provided of course that the Higgs has a lower mass than the Z) or any of the other new types of particle discussed in chapter 13, as well as producing clean quark and gluon hadronic jets of very high energy. Higher energy ep collisions to probe the deep structure of the proton will also be possible when the HERA colliding beam facility (30 GeV electrons on 820 GeV protons) is constructed in Hamburg.

In this book we have tried to provide a fairly simple survey of hadronic processes from a modern parton viewpoint, stressing the relationship of the QCD-based parton model of hadron structure to the Regge theory of soft hadronic interactions on the one hand, and to the electroweak flavour dynamics on the other. The key to the understanding of hadron reaction mechanisms lies, it would seem, in the marriage of these approaches.

REFERENCES

Abarbanel H D I, Bronzan J B, Sugar R L and White A R 1975 *Phys. Rep.* **21C** 120
Abers E and Lee B W 1973 *Phys. Rep.* **9C** 1
AFS collaboration 1982 *Phys. Lett.* **118B** 185
—— 1983a *Phys. Lett.* **128B** 354
—— 1983b *Phys. Lett.* **123B** 367
Aguilar-Benitez M *et al* 1983 *Phys. Lett.* **123B** 98
Aitchison I J R and Hey A J G 1982 *Gauge Theories in Particle Physics* (Bristol: Adam Hilger).
Aitkenhead W *et al* 1980 *Phys. Rev. Lett.* **45** 157
Albrecht A and Steinhardt P J 1982 *Phys. Rev. Lett.* **48** 1220
Albrow M G *et al* 1978 *Nucl. Phys.* B **145** 305
Altarelli G 1982 *Phys. Rep.* **81C** 1
Altarelli G, Ellis R K and Martinelli G 1979 *Nucl. Phys.* B **157** 461
Altarelli G and Parisi G 1977 *Nucl. Phys.* B **126** 298
Amaldi U 1979 *CERN Yellow Rep.* 79-06
Amaldi U *et al* 1977 *Phys. Lett.* **66B** 390
Amati D and Veneziano G 1979 *Phys. Lett.* **83B** 87
Andersson B, Gustafson G and Peterson C P 1977 *Phys. Lett.* **69B** 221, **71B** 337
Andersson B, Gustafson G and Sjöstrand T 1980 *Z. Phys.* C **6** 235, *Phys. Lett.* **94B** 211
Andersson B, Gustafson G and Soderberg B 1983 *Z. Phys.* to be published
Anisovich V V and Shekhter V M 1973 *Nucl. Phys.* B **55** 455
Antoniou N G *et al* 1983 *Phys. Lett.* **128B** 257
Asa'd Z *et al* 1983 *Phys. Lett.* **128B** 124, **130B** 335
Baier R and Rückl R 1983 *Z. Phys.* C **19** 251
Bailin D 1982 *Weak Interactions* (Bristol: Adam Hilger)
Barger V 1974 *Proc. 17th Conf. on High Energy Physics, London* ed J R Smith (Didcot: Rutherford Laboratory) I-193
Barger V, Baer H, Martin A D, Glover E W N, Phillips R J N 1984a *Phys. Rev.* to be published
Barger V, Baer H, Martin A D and Phillips R J N 1984b *Phys. Rev.* D **29** 887
Barger V, Halzen F and Keung W Y 1982 *Phys. Rev.* D **25** 112
Barger V, Martin A D and Phillips R J N 1983a *Phys. Lett.* **125B** 339
—— 1983b *Phys. Lett.* **125B** 343
—— 1983c *Z. Phys.* C **21** 99
Barger V and Phillips R J N 1974 *Nucl. Phys.* B **73** 269
Barnes A V *et al* 1976 *Phys. Rev. Lett.* **37** 76
—— 1978 *Nucl. Phys.* B **145** 67
Bartel W *et al* 1983 *Phys. Lett.* **123B** 353
Bartels J 1980 *Nucl. Phys.* B **175** 365
Basile M *et al* 1980 *Phys. Lett.* **92B** 367
—— 1981 *Phys. Lett.* **99B** 247
Basile M *et al* 1984 *Nuovo Cimento* **79A** 1
Bassetto A, Ciafaloni M and Marchesini G 1980 *Nucl. Phys.* B **163** 477
—— 1984 *Phys. Rep.* **100C** 201
Benary O, Gotsman E and Lissauer D 1983 *Z. Phys.* C **16** 211
Benecke J, Chou T T, Yang C N and Yen E 1969 *Phys. Rev.* **188** 2159
Berge J P *et al* 1980 *Phys. Lett.* **91B** 311
Berger Ch *et al* 1980 *Phys. Lett.* **95B** 313
Berkelman K 1983 *Phys. Rep.* **98C** 146
Bilenky S M and Hosek J 1982 *Phys. Rep.* **90C** 73
Bionta R M *et al* 1983 *Phys. Rev. Lett.* **51** 27
Bjorken J D 1973 *Proc. 1973 SLAC Summer Inst.* SLAC-167

—— 1978 *Phys. Rev.* D **17** 171

Bjorken J D and Brodsky S J 1970 *Phys. Rev.* D **1** 1416

Bjorken J D and Drell S D 1964 *Relativistic Quantum Mechanics* (New York: McGraw-Hill)

Blankenbecler R and Brodsky S J 1974 *Phys. Rev.* D **10** 2973

Blankenbecler R, Brodsky S J and Gunion J 1975 *Phys. Rev.* D **12** 3469

Blankenbecler R, Brodsky S J, Gunion J F and Savit R 1973 *Phys. Rev.* D **8** 4117

—— 1974 *Phys. Rev.* D **10** 2153

Bobbink G J *et al* 1980 *Phys. Rev. Lett.* **44** 118

Bohm A 1980 *Proc. 20th Int. Conf. on High Energy Physics, Madison* ed L Durand and L G Pondrom *AIP Conf. Proc.* **68** 551

Breakstone A *et al* 1983 *Phys. Lett.* **132B** 458; **132B** 463

Brodsky S J 1979a *Phys. Scr.* **19** 154

—— 1979b *Proc. 1979 SLAC Summer Inst. on Quantum Chromodynamics* ed A Mosher, p 133

Brodsky S J and Drell S D 1980 *Phys. Rev.* D **22** 2236

Brodsky S J and Farrar G 1973 *Phys. Rev. Lett.* **31** 1153

Brodsky S J and Gunion J F 1976 *Phys. Rev. Lett.* **37** 402

Brodsky S J, Hoyer P, Peterson C and Sakai N 1981 *Phys. Lett.* **93B** 451

Brodsky S J and Lepage P 1979a *Phys. Rev. Lett.* **43** 545

—— 1979b *Phys. Lett.* **87B** 359

Brodsky S J and Weiss N 1977 *Phys. Rev.* D **16** 2325

Buras A J 1980 *Rev. Mod. Phys.* **52** 199

Buras A J and Gaemers K J F 1978 *Nucl. Phys.* B **132** 249

Capella A 1981 *Proc. Europhysics Study Conf. on Partons in Soft Hadronic Processes, Erice*

Capella A and Krzywicki A 1983 *Orsay Prepr.* LPTHE 83/12

Capella A, Sukhatme U, Tan C I and Tran Thanh Van J 1979 *Phys. Lett.* **81B** 68, **93B** 146

Capella A, Sukhatme U and Tran Thanh Van J 1980 *Z. Phys.* C **3** 329

Carlitz R, Green M B and Zee A 1971 *Phys. Rev.* D **4** 3439

Carroll A S *et al* 1976 *Phys. Lett.* **61B** 303

Chamseddine A H, Nath P and Arnowitt R 1983 *Phys. Lett.* **129B** 445

CHARM collaboration 1981 *Phys. Lett.* **99B** 265

Chau L L 1983 *Phys. Rep.* **95C** 1

Chauvat P *et al* 1983 *Phys. Lett.* **127B** 384

Cheng H, Walker J K and Wu T T 1973 *Phys. Lett.* **44B** 283

Chou T T and Yang C N 1968 *Phys. Rev.* **170** 5

—— 1983 *Phys. Lett.* **128B** 457

Close F 1979 *An Introduction to Quarks and Partons* (New York: Academic)

—— 1982 *Phys. Scr.* **25** 86

—— 1983 *Proc. High Energy Phys. Conf. Brighton* p 361

Cohen-Tannoudji G *et al* 1980 *Phys. Rev.* D **21** 2699

Collins P D B 1977 *Regge Theory and High Energy Physics* (Cambridge: Cambridge University Press)

Collins P D B and Gault F D 1978 *Phys. Lett.* **73B** 330

—— 1982 *Phys. Lett.* **112B** 255

Collins P D B, Gault F D and Martin A 1974 *Nucl. Phys.* B **83** 241

Collins P D B, Johnson R C and Squires E J 1968 *Phys. Lett.* **26B** 223

Collins P D B and Kearney P J 1983 *Z. Phys.* C to be published

Collins P D B and Wilkie T D B 1981 *Z. Phys.* C **7** 357

Collins P D B and Wright A D M 1979 *J. Phys. G: Nucl. Phys.* **5** 1461

Combridge B L 1979 *Nucl. Phys.* B **151** 429

Combridge B L, Kripfganz J and Ranft J 1977 *Phys. Lett.* **70B** 234

Combridge B L and Maxwell C 1984 *Rutherford Laboratory Rep.* RL-83-095 to be published

Cool R L *et al* 1981 *Phys. Rev. Lett.* **47** 701

Coon D D, Gunion J F, Tran Thanh Van J and Blankenbecler R 1978 *Phys. Rev.* D **18** 1451

Creutz M 1984 *Quarks, Gluons and Lattices* (Cambridge: Cambridge University Press)

Cutler R and Sivers D 1978 *Phys. Rev.* D **17** 196

Darriulat P 1980 *Ann. Rev. Nucl. Particle Sci.* **30** 159

Das K P and Hwa R C 1977 *Phys. Lett.* **68B** 459

Denegri D *et al* 1981 *Phys. Lett.* **98B** 127

De Rujula A, Ellis J, Floratos E G and Gaillard M K 1978 *Nucl. Phys.* B **138** 387

De Tar C E 1971 *Phys. Rev.* D **3** 128

Dias de Deus J and Kroll P 1978 *Acta Phys. Pol.* B **9** 159

Dicus D A, Nandi S, Repko W and Tata X 1984 *Phys. Rev.* D **29** 67

Di Lella L 1979 *Proc. 10th Symp. on Multiparticle Dynamics*

Dokshitzer Y L, D'Yakanov D I and Troyan S I 1980 *Phys. Rep.* **58C** 271

Dolgov A D and Zeldovich Ya B 1981 *Rev. Mod. Phys.* **53** 1

Donnachie A and Landshoff P V 1979 *Z. Phys.* C **2** 55

—— 1983 *Phys. Lett.* **123B** 345

—— 1984 *Nucl. Phys.* B **231** 189

Drell S D and Yan T M 1970 *Phys. Rev. Lett.* **24** 181

—— 1971 *Ann. Phys., NY* **66** 578

Duke D W and Taylor F E 1978 *Phys. Rev.* D **17** 1788

Eden R J, Landshoff P V, Olive D I and Polkinghorne J C 1966 *The Analytic S Matrix* (Cambridge: Cambridge University Press)

Eichten E *et al* 1980 *Phys. Rev.* D **21** 203

Ellis J 1982 *Phil Trans. R. Soc.* A **307** 21

Ellis J, Gaillard M K and Nanopoulous D V 1976a *Nucl. Phys.* B **106** 292

Ellis J, Gaillard M K and Ross G G 1976b *Nucl. Phys.* B **111** 253

Ellis J, Hagelin J S, Nanopoulous D V and Srednicki M 1983 *Phys. Lett.* **127B** 233

Ellis J, Nanopoulos D V and Rudaz S 1982 *Nucl. Phys.* B **202** 43

Ellis J and Sachrajda C T 1979 *Quarks and Leptons, Cargese 1979. NATO Advanced Study Institutes* B **61** 285

Ellis S D, Fleishon N and Stirling W J 1981 *Phys Rev.* D **24** 1386

Farhi E 1977 *Phys. Rev. Lett.* **39** 1587

Farrar G and Neri F 1983 *Phys. Lett.* **130B** 109

Fayet P and Ferrara S 1977 *Phys. Rep.* **32C** 250

Feynman R P 1969 *Phys. Rev. Lett.* **23** 1415

—— 1972 *Photon–Hadron Interactions* (New York: Benjamin)

Feynman R P and Field R D 1977 *Phys. Rev.* D **15** 2590

—— 1978 *Nucl. Phys.* B **136** 1

Feynman R P, Field R D and Fox G C 1978 *Phys. Rev.* D **18** 3320

Fiałkowski K and Kittel W 1983 *Rep. Prog. Phys.* **46** 1283

Field R D 1979 *Phys. Scr.* **19** 131, *AIP Conf. Proc.* **55** 97

Field R D and Fox G C 1974 *Nucl. Phys.* B **80** 367

Finjord J, Girardi G and Sorba P 1979 *Phys. Lett.* **89B** 99

Fox G C 1977 *Nucl. Phys.* B **131** 107

Fritzsch H and Minkowski P 1981 *Phys. Rep.* **73C** 67

Froissart M 1961 *Phys. Rev.* **123** 1053

Furmanski W, Petronzio R and Pokorski S 1979 *Nucl. Phys.* B **155** 253

Ganguli S N and Roy D P 1980 *Phys. Rep.* **67C** 201

Gasser J and Leutwyler H 1982 *Phys. Rep.* **87C** 77

Gell-Mann M 1961: reprinted in Gell-Mann M and Ne'eman Y 1964 *The Eightfold Way* (New York: Benjamin)

—— 1964 *Phys. Lett.* **8** 214

Giacomelli G and Jacob M 1979 *Phys Rep.* **55C** 1

Gilman F J 1972 *Phys. Rep.* **4C** 95

Glashow S L 1961 *Nucl. Phys.* **22** 579

Glover E W N, Halzen F, Herzog F and Martin A D 1984 *Durham Prepr.* 84/8

Goldberg H 1972 *Nucl. Phys.* B **44** 149

Gourdin M 1974 *Phys. Rep.* **11C** 29

Greenberg O W and Nelson C A 1977 *Phys. Rep.* **32C** 1

Grisaru M T, Schnitzer H J and Tsao H S 1973 *Phys. Rev.* D **8** 4498

Gross D J and Wilczek F 1973 *Phys. Rev.* D **8** 3633

Gunion J F 1979 *Phys. Lett.* **88B** 150

—— 1980 *Proc. 11th Int. Symp. on Multi-Particle Dynamics, Bruges, Belgium*

Guth A 1981 *Phys. Rev.* D **23** 347

Halliday I G 1983 *Proc. High Energy Phys. Conf. Brighton* p 506

—— 1984 *Rep. Prog. Phys.* to be published

Halzen F 1982 *Proc. XXI Conf. on High Energy Physics, Paris*: *J. Physique* **C3** 381

Halzen F and Hoyer P 1983 *Phys. Lett.* **130B** 326

Halzen F and Martin A D 1984 *Quarks and Leptons* (New York: Wiley)

Halzen F, Martin A D and Scott D M 1982a *Phys. Lett.* **112B** 160

Halzen F, Martin A D, Scott D M and Tuite M P 1982b *Z. Phys.* C **14** 351

Halzen F and Mursula K 1983 *Phys. Rev. Lett.* **51** 857

Halzen F and Scott D M 1978 *Phys. Rev. Lett.* **40** 1117

Hara T *et al* 1983 *Phys. Rev. Lett.* **50** 2058

Harari H 1978 *Phys. Rep.* **42C** 235

—— 1980 *Proc. 1980 Summer Inst.* ed A Mosher (Stanford: SLAC) p 141

Harari H and Seiberg N 1981 *Phys. Lett.* **98B** 269

—— 1982 *Nucl. Phys.* B **204** 141

Hey A J G and Morgan D 1978 *Rep. Prog. Phys.* **41** 675

Hikasa K 1984 *Univ. Wisconsin Rep.* MAD/PH/144

Horgan R and Jacob M 1981 *Phys. Lett.* **107B** 395

Hoyer P *et al* 1979 *Nucl. Phys.* B **161** 349

Hwa R C 1980 *Phys. Rev.* D **22** 759, 1593

Hwa R C and Roberts R G 1979 *Z. Phys.* C **1** 81

Hwa R C and Zahir M S 1981 *Phys. Rev.* D **23** 2539

Iliopolous J 1976 *CERN Yellow Rep.* 76-11

Irving A C and Worden R 1977 *Phys. Rep.* **34C** 117

Isgur N 1978 in *Proc. XVI Int. School of Subnucl. Phys. Erice* ed A Zichichi (New York: Plenum)

Isgur N and Karl G 1978 *Phys. Rev.* D **18** 4187

—— 1979 *Phys. Rev.* D **20** 1191

Itzykson C and Zuber J-B 1980 *Quantum Field Theory* (New York: McGraw-Hill)

Jacob M (ed) 1974 *Dual Theory* (Amsterdam: North Holland)

—— 1983a in *Proc. 3rd Topical Workshop on Proton-antiproton Collider Physics* CERN 83-04 ed C Bacci and G Salvini

—— 1983b *Collider Physics—present and prospects, SLAC Summer Institute proceedings 1983*

Jacob M and Landshoff P V 1978 *Phys. Rep.* **48C** 285

Jaffe R and Johnson K 1975 *Phys. Lett.* **60B** 201

Jauch J M and Rohrlich F 1955 *Theory of Photons and Electrons* (New York: Academic)

Jones D and Gunion J F 1979 *Phys. Rev.* D **19** 867

Kane G L and Seidl A 1976 *Rev. Mod. Phys.* **48** 309

Kanwal S 1983 *PhD Thesis* 'A leading order QCD computation of $\pi\pi$ scattering' (Calif. Inst. Tech.)

Keung W Y 1982 *AIP Conf. Proc.* **85** 186

Kittel W 1981 *Proc. Europhysics Study Conf. on Partons in Soft Hadronic Processes, Erice* ed R T Van de Walle (Singapore: World Scientific) p 1

Kleinknecht K and Renk B 1983 *Phys. Lett.* **130B** 459

Kogut J and Susskind L 1974 *Phys. Rev.* D **9** 697, 3391

Kokkedee J J J 1969 *The Quark Model* (New York: Benjamin)

Kolb E W and Turner M S 1983 *Ann. Rev. Nucl. Part. Phys.* **33** 645

Konishi K, Ukawa A and Veneziano G 1979 *Nucl. Phys.* B **157** 45

Kowalski H 1983 *Proc. High Energy Phys. Conf. Brighton* p 152

Landshoff P V 1974 *Phys. Rev.* D **10** 1024

—— 1983 *Small p_T Physics: DAMTP report*

Langacker P 1981 *Phys. Rep.* **72C** 185

Lattes C, Fujimoto Y and Hasegawa S 1980 *Phys. Rep.* **65C** 151

Leader E and Predazzi E 1982 *Gauge Theories and the New Physics* (Cambridge: Cambridge University Press)

Lifshitz E and Pitayevski 1976 *Relativistic Quantum Mechanics* (Oxford: Pergamon)

Linde A D 1982 *Phys. Lett.* **108B** 389

Llewellyn-Smith C H 1972 *Phys. Rep.* **3C** 264

Low F E 1975 *Phys. Rev.* D **12** 163

Lyons L 1983 *Prog. Part. and Nucl. Phys.* **10** 227

McCubbin N A 1981 *Rep. Prog. Phys.* **44** 1027

Magnus W and Oberhettinger F 1949 *Functions of Mathematical Physics* (New York: Chelsea) p 4

Marciano W and Pagels H 1978 *Phys. Rep.* **36C** 1

Marciano W and Sirlin A 1981 *Nucl. Phys.* B **189** 442

Martin A 1963 *Strong Interactions and High Energy Physics* ed R G Moorhouse (Edinburgh: Oliver and Boyd)

—— 1966 *Nuovo Cimento* **42** 930, **44** 1219

—— 1983 *Proc. 3rd Topical Workshop on Proton–antiproton Collider Physics* CERN 83-04 ed C Bacci and G Salvini p 351

Matthiae G 1983 *Proc. High Energy Phys. Conf. Brighton* p 714

Morse P M and Feshbach H 1953 *Methods of Theoretical Physics* (New York: McGraw-Hill)

Moshe M 1978 *Phys. Rep.* **37C** 255

Mueller A H 1970 *Phys. Rev.* D **2** 2963

—— 1981 *Phys. Rep.* **73C** 237

Muller F 1983 *CERN Rep.* EP 83-67

Ne'eman Y 1961 *Nucl. Phys.* **26** 222

van Nieuwenhuizen P 1981 *Phys. Rep.* **68C** 189

Nussinov S 1976 *Phys. Rev. Lett.* **34** 1286

Ochs W 1977 *Nucl. Phys.* B **118** 397

Odorico R 1982 *AIP Conf. Proc.* **85** 100

—— 1983 *CERN Prepr.* TH 3662

Okun L B 1982 *Leptons and Quarks* (Amsterdam: North Holland)

Ono S and Schoberl F 1982 *Phys. Lett.* **118B** 419

Owens J F, Reya E and Gluck M 1978 *Phys. Rev.* D **18** 1501

Pakvasa S, Dechantsreiter M, Halzen F and Scott D M 1979 *Phys. Rev.* D **20** 2862

Parisi G 1980 *Phys. Lett.* **90B** 295

Parisi G and Petronzio R 1979 *Nucl. Phys.* B **154** 427

Particle Data Group 1982 *Phys. Lett.* **111B** 1

Pennington M R 1983 *Rep. Prog. Phys.* **46** 293

Perl M L 1974 *High Energy Hadron Physics* (New York: Wiley)

Perl M L *et al* 1975 *Phys. Rev. Lett.* **35** 1489

Peterson C 1979 *Proc. Topical Workshop on Forward Production of High-mass Flavours at Collider Energies, College de France, Paris*

Peterson C, Schlatter D, Schmitt I and Zerwas P M 1983 *Phys. Rev.* D **27** 105

Pokorski S and Van Hove L 1975 *Nucl. Phys.* B **86** 243

Politzer H D 1973 *Phys. Rev. Lett.* **30** 1346

—— 1974 *Phys. Rep.* **14C** 130

Polkinghorne J C 1980 *Models of High Energy Processes* (Cambridge: Cambridge University Press)

Pomeranchuk I Y 1958 *Sov. Phys.–JETP* **7** 499

Quigg C and Rosner J L 1979 *Phys. Rep.* **56C** 167

Rakow P E L and Webber B R 1981 *Nucl. Phys.* B **187** 254

Regge T 1959 *Nuovo Cimento* **14** 951

—— 1960 *Nuovo Cimento* **18** 947

Reya E 1981 *Phys. Rep.* **69C** 195

Roberts R G and Roy D P 1974 *Nucl. Phys.* B **77** 240

Rosner J L 1969 *Phys. Rev. Lett.* **22** 689

Ross G G 1981 *Rep. Prog. Phys.* **44** 655

Rubbia C 1983 *Proc. High Energy Phys. Conf. Brighton* p 860

—— 1984 *4th Topical Workshop on Proton–antiproton Collider Physics, Bern*

Rubbia C, McIntyre P and Cline D 1976 *Proc. Int. Neutrino Conference Aachen 1976* (Braunschweig: Vieweg) p 683

Rushbrooke J G and Webber B R 1978 *Phys. Rep.* **44C** 1

Salam A 1968 *Proc. 8th Nobel Symp.* ed N Svartholm (Stockholm: Amqvist and Wiksell)

Scott D M 1978 *Phys. Rev.* D **18** 210

SFM collaboration 1984 *Phys. Lett.* **135B** 505, 510

Sivers D, Brodsky S J and Blankenbecler R 1976 *Phys. Rep.* **23C** 1

Söding P 1983 *Proc. High Energy Phys. Conf. Brighton* p 567

Sommerfeld A 1949 *Partial Differential Equations in Physics* (New York: Academic)

Sosnowski R 1983 *Proc. High Energy Phys. Conf. Brighton* p 628

Steigman G 1979 *Ann. Rev. Nucl. Sci.* **29** 313

Sterman G and Weinberg S 1977 *Phys. Rev. Lett.* **39** 1436

Stroynowski R 1981 *Phys. Rep.* **71C** 1

Suzuki M 1977 *Phys. Lett.* **71B** 139

TASSO collaboration 1983 *Z. Phys.* C **17** 5

UA1 collaboration 1983a *Phys. Lett.* **132B** 223

—— 1983b *Phys. Lett.* **123B** 115

—— 1983c *Phys. Lett.* **132B** 214

—— 1983d *Phys. Lett.* **122B** 103

—— 1983e *Phys. Lett.* **129B** 273

—— 1983f *Phys. Lett.* **126B** 398

UA2 collaboration 1982 *Phys. Lett.* **118B** 203

—— 1983a *Z. Phys.* C **20** 117

—— 1983b *Phys. Lett.* **122B** 476

—— 1983c *Phys. Lett.* **129B** 130

UA4 collaboration 1983 *Proc. High Energy Phys. Conf. Brighton* p 724

UA5 collaboration 1981 *Phys. Lett.* **107B** 315

Van Hove L 1979 *Proc. 18th Int. Universitätswochen für Kernphysik, Schladming*

—— 1982 *Phys. Lett.* **127B** 138

Veneziano G 1968 *Nuovo Cimento* A **57** 190

Watson G N 1918 *Proc. R. Soc.* **95** 83

Webber B R 1982 *Phys. Scr.* **25** 198

Wegener D 1983 *Proc. High Energy Phys. Conf. Brighton* p 121

Weinberg S 1967 *Phys. Rev. Lett.* **19** 1264

—— 1977 *The First Three Minutes* (New York: Basic)

West G B 1970 *Phys. Rev. Lett.* **24** 181

Wheater J F and Llewellyn Smith C H 1982 *Nucl. Phys.* B **208** 189

White A R 1980 *CERN Prepr.* TH 2976

—— 1981 *AIP Conf. Proc.* **85** 363

—— 1983 *Ann. Phys., NY* in press

Wiik B H 1980 *Proc. 20th Int. Conf. High Energy Physics, Madison* ed L Durand and L G Pondrom *AIP Conf. Proc.* **68** 1379

Wiik B H and Wolf G 1979 *Electron–positron Interactions: Springer Tracts in Modern Physics* **86** (Berlin: Springer-Verlag)

Wilczek F 1982 *Ann. Rev. Nucl. Part. Sci.* **32** 177

Wolf G 1980 *Proc. 11th Int. Symp. Multiparticle Dynamics, Bruges*

—— 1982 *Proc. 21st Int. Conf. on High Energy Physics, Paris: J. Physique* **C3** 525

Yang C N and Mills R L 1954 *Phys. Rev.* **96** 191

Zweig G 1964 *CERN Rep.* TH 401, 412 unpublished

INDEX

Note. P.8

as long as $N_f < 16$ α_s runs "down"

P.39. one source of scaling violation.